# Cases in Corporate Sustainability and Change

## A Multidisciplinary Approach

**Suzanne Benn, Dexter Dunphy and Bruce Perrott**

Cases in Corporate Sustainability and Change: A Multidisciplinary Approach
1st edition, 1st printing

**Author**
Suzanne Benn, Dexter Dunphy and Bruce Perrott

**Cover designer**
Christopher Besley, Besley Design

**ISBN**: 978-0-7346-1128-4

**Disclaimer**

All reasonable efforts have been made to ensure the quality and accuracy of this publication. Tilde University Press assumes no responsibility for any errors or omissions and no warranties are made with regard to this publication. Neither Tilde University Press nor any authorised distributors shall be held responsible for any direct, incidental or consequential damages resulting from the use of this publication.

Published in Australia by:
**Tilde University Press**
**PO Box 72**
**Prahran VIC 3181 Australia**
www.tup.net.au

# CONTENTS

# PREFACE

**Why this book now?** Today humankind faces two major crises. The first is the global financial crisis (GFC) which began in the second half of 2008 and was initiated by the near meltdown of the world's financial system. The second is the ecological crisis, particularly climate change – a crisis that has been slowly building since the industrial revolution. The ecological crisis is now gaining momentum as we witness the meltdown of the world's glaciers, permafrost and sea ice – and a range of related issues such as widespread weather volatility, desertification and food shortages. The governments of the world's leading nation states acted quickly and decisively to reduce the impact of the GFC but have been unable or unwilling to establish agreement for action on the ecological crisis.

The two crises are intimately related. In a perceptive account, Thomas Friedman explains "why the Market and Mother Nature hit the wall at the same time and for the same reasons" (Friedman 2009, p. 8). Friedman argues that our current values, expressed in our lifestyle, are creating toxic assets in both the financial world and the natural world and that these toxic assets are overwhelming us. In his view, the reasons for this are that we systematically underprice the true costs and risks of what we do, maximise short term returns and so fail to manage for the future, and privatise gains while socialising losses. Friedman is only one of many authors now arguing that we need a fundamental transformation of how we live and work in order to avoid a cataclysmic collapse of human civilisation (Flannery 2009; Flannery 2010; Hamilton 2010; Lovelock 2009; Monbiot 2006; O'Brien 2009; Spratt & Sutton 2008). In the words of leading ecologist Tim Flannery: we have become the "future eaters", living beyond the earth's ability to replace the resources we consume (Flannery 2002).

Nevertheless, change is occurring in these characteristics of the economy that are a carryover from commitment to a 'free market economy' and ever expanding consumption – the ideology of the last century. In one sense the challenge we face is unique. Never before in history has humankind become the most powerful force affecting the planet; the combination of growing population and powerful technologies is novel and unprecedented. In another sense however the situation is not novel. This is not the first time in human history that a major economic, social and political transformation has occurred. For example, the transition from hunting and gathering bands to an agrarian society was such a change. Similarly the transition from an agrarian society to an industrial, carbon-based society – the industrial revolution – was also such a change. Our survival now depends on making a similar transition – this time moving from a carbon-based economy to a carbon neutral economy using alternate technologies such as solar, wind and thermal power. The revolution is already underway and this book charts some of the steps that organisations are making to contribute to the transformation.

Organisations are the cells of modern society, component units of the economy, and they are already changing in ways that can inform us about the nature of the new world that is emerging. More and more organisational leaders are concerned about social and environmental issues and see the positive benefits of eliminating the extraordinary waste that characterises much of contemporary organisational life as well as anticipating the business benefits of actively shaping the future instead of reacting after the event. Of course tackling the issues identified above is an ethical imperative, but it also is opens up a potentially better way of doing business that brings benefits to stakeholders and, in the case of public companies, also to shareholders. What is increasingly apparent to perceptive executives is that business as usual is no longer an option, that post GFC, the world is not going to revert to the old 'stable state' but rather tip forward into a new world that we do not yet fully understand and cannot yet specifically delineate. In times of major transitions, trying to anticipate an unknown future is risky, but clinging to the delusional security of past practices is the greatest risk of all. No organisation can assume that it has a place in the future – the price of survival now is constant innovation based on effective scanning of emerging trends and discontinuities.

There is a growing awareness and acceptance in society and in the business community of the need to create sustainable and sustaining organisations. Signs of this are readily discernible in the increasing role being accorded sustainability in conference agendas and in the exponential increase in publications on sustainability, including new journals devoted to the area. The trend is also evident in new specialist business school courses in sustainability and an increasing emphasis on sustainability across the range of traditional courses throughout school, university and technical and further education curricula. Most importantly, there are widespread business initiatives, emerging political debates and hotly debated government legislation in a range of areas related to sustainability such as carbon emissions, water allocation and alternative energy technologies, as well as non-technological areas such as corporate governance, occupational health and safety and equal opportunity.

The shift in awareness is much greater than many recognise. According to an Accenture Report for the 2010 UN Global Compact, 93% of the 766 CEOs of global companies surveyed in 2010 believed that sustainability issues will be critical to the future success of their organisations, and 96% believed that sustainability issues should be fully integrated into strategy and operations. These figures show a dramatic rise from a similar survey undertaken in 2007. Rather than the economic downturn lessening the importance of sustainability, the CEOs said the downturn has raised the importance of sustainability<www.unglobalcompact.org/news/42-06-22-2010>. We, the editors of this book, are researchers and consultants participating in corporate life. Our own experience provides evidence of these changing attitudes and also that attitudinal change is translating into action. Sustainability is emerging as the central theme for the future definition of the twenty-first century world. This book is a contribution to the emerging dialogue about the nature of that world. Fortunately, the dialogue around sustainability is no longer being conducted in words alone but in unprecedented experimentation with new organisational forms and practices.

# The centrality of organisational change

As the title of this book indicates our primary emphasis is on corporate sustainability, but we have also placed emphasis on corporate change. The two are inextricably linked, for making organisations sustainable represents a major transformation of how in the past they have conducted their business. Sustainability will not be achieved through technological fixes alone; while the technological base of society and organisations needs to change dramatically, so does the corporate culture which is needed to deploy the new technologies effectively. The urgency of the changes also demands that the transformation of organisations occurs rapidly and this requires a new level of change management competency beyond the current level of most contemporary organisations.

Every two years IBM conducts a survey of more than 1,000 CEOs of global companies. The last two surveys, made in 2008 and 2010, have shown the importance of building a change management competency into modern organisations. The 2008 survey was conducted before the GFC and focussed on asking CEOs to outline what they believed would be key characteristics of the successful enterprise of the future (IBM Corporation 2008). Among the significant characteristics they identified for future-fit organisations were that they would be: *hungry for change*, *innovative beyond customer imagination* and *disruptive by nature*. These CEOs of some of the world's largest and most successful organisations saw the need to go beyond adapting to change, to seizing leadership in actively shaping the future. They suggested that the enterprise of the future would be home to 'visionary challengers', that is, people who question assumptions and suggest radical alternatives. Eight out of ten CEOs expected substantial change over the next three years yet they believed that the capacity of their organisation to manage change was well below the level needed for success.

The issues identified in the 2008 study were followed up in another IBM study *Making Change Work* (IBM Corporation 2008). This time the interviewees were not the CEOs but 1500 managers of major projects in global companies, projects that involved introducing major change. Many of these projects were multi-million dollar projects. The project leaders reported that only 14% of projects were considered successful, that is, met their objectives against time, budget and quality goals. Forty-four per cent of projects failed to meet these criteria in some way and 15% failed completely. These shortfalls represent, amongst other consequences, huge cost overruns. However the most successful organisations (the top 20% in terms of project success) had an 80% success rate, nearly double the average and 20 times the bottom 20% of organisations studied.

Clearly there are many organisations that fail to manage large scale transitions. However a sizeable proportion of organisations do succeed in embedding the skills needed for transformational culture change and in deploying those skills effectively. We need case studies of how these 'change masters' (as IBM calls them) manage major transitions and transformations. This book includes examples from which we can learn.

But managing rapid change is only one major challenge for modern managers. Another related challenge is the increasing complexity involved in managing change when environments are not only changing fast, and often unpredictably, but also complexifying. The most recent IBM Global CEO study, *Capitalising on Complexity* (IBM Corporation 2010) interviewed over 1500 global CEOs and adds a new challenge to the earlier emphasis on the need to increase investment in change capabilities. This challenge arises from the need to manage increasing connectivity and the resulting strong interdependencies. The new business environment is not only more uncertain and more volatile; it is also more complex as organisations move toward managing from the outside in. Six out of ten CEOs reported that they regard the new environment as more complex and that they expect complexity to increase in future. In our view the move to sustainability adds to this complexity because managers of private sector organisations charged with implementing sustainability have to respond to a much wider range of stakeholders, not just to shareholders, and need to understand a range of non-market factors such as environmental, social and workforce trends. Sustainability requires bringing together scientific or technical experts with employees from across the business, integrating a wide range of disciplinary and functional backgrounds and their knowledge bases. For managers to advance sustainability performance they need to recognise the interconnectedness and complexity of sources of environmental impact and their relationships to social and economic concerns of the organisation. In short, they must address the systemic and holistic nature of sustainability knowledge (Porter 2008). Hence the move to create sustainable organisations by no means simplifies the managerial task. In this book, we have examples of managers working on coming to grips with this new level of complexity.

The high performance organisations of the future will create cultures that deal effectively with both system-level change and complexity (Doppelt 2009; Galea 2009; Jones *et al.* 2011). In these case studies we can see this new level of competence emerging. For example, the sustainability change process described in the Yarra Valley Water case involves linking different levels of systems thinking – the scientific and functional level with the stakeholder interactive and complex adaptive systems levels (Porter & Cordoba 2009). The case study involving the City of Mandurah local government agency shows us that both enabling and transformational leadership styles as well as a range of key skills and attributes are associated with effective sustainability change programs.

## Why cases?

This is a case book. This prompts the question of why we decided to produce a book of this kind and not, for example, another contribution to managerial theory or a textbook on new approaches to sustainability.

Let us stress at this point that we are very enthusiastic about theory. Between us we have published several books of organisational theory. As Kurt Lewin once said: "There is nothing more practical than a good theory". Without theory it is impossible to generalise from experience – we are doomed to simply repeat past failures or act randomly in the hope we may do better next time without knowing, even if we succeed, why we did so. There is in fact a good deal of theory in this book.

But theory divorced from experience and practice becomes so rarefied as to be largely useless; simply an intellectual exercise. The best test of a theory is its ability to interpret experience, to explain relationships between key variables and predict outcomes. Professions test theory in action and codify best practice into theory. We have chosen in this book to select authors who are involved, as we are, in studying organisations where managers are actively exploring how to make their organisations more sustainable.

The path to sustainability is not fully defined; if we use the analogy of crossing a river, we are crossing the river by feeling for a footing on the stones on the river bed, and some are trying at one spot along the river and others at another. So the cases we have gathered together are diverse. They are drawn from different industries; they are from the private sector, the public sector and the not-for-profit sector; they are large, medium and small organisations. What they have in common is a willingness of at least some in these organisations, to try new things, to innovate and to question the way things have been traditionally done. These change agents are the new heroes of the twenty-first century.

This innovation, this risk taking, has a definite direction – it represents a step toward making the organisation *in some sense* more sustainable. There is by no means agreement across the organisations presented here about what sustainability means; even within these organisations there is debate about that. That's the reality – we are still exploring what sustainability means in any situation, trying new things, looking for something that gives traction in our move away from the waste and destruction of the natural environment, of resources, of people and toward something that carries us forward to a more sane and truly productive world of caring for the planet, for each other, for other species.

So the cases here are about sustainability and change and particularly about the leadership of change. Leadership after all is about *stepping out*. As Kahane (2010, p. 116) writes: "To lead means to step forward, to exceed one's authority, to try to change the status quo, to exercise power, and such action is by definition disruptive. There is no way to change the status quo without discomforting those who are comfortable with the status quo". These cases show sustainability leaders trying new initiatives at work.

So why *cases*?

Constructing a viable path to a more sustainable world means documenting our stepping out, our stumbling forward, our novel ideas, our successes and failures, our learnings. Cases can provide insights into the formation of the future; they can provide a basis for generalisation about how the future may be shaping up, and insights about what the future world of work may look like, at least in part.

A lot of what is written about sustainability concerns the dire situation in which we find ourselves and the imminent disasters that will eventuate unless we mend our ways. But if we want the world to change, we can learn from the words of the creative genius Buckminster Fuller who said: "You never change things by fighting existing reality. To change something, build a new model that makes the existing model obsolete" (quoted in Kahane 2010, pp. 112–113). Case studies provide the basis for a new model to supplant the old. However they do not come to us with a fully formed new model. To get a real fix on the future, we have to extrapolate from a number of cases – such as those we have gathered together in this book. Individual cases are like pieces of a jigsaw puzzle that must be fitted together to make a meaningful picture. In one organisation we find a new way to understand how to remanufacture, in another a way to recycle used products, in another a way to build an innovative culture that stimulates sustainability initiatives, and in yet another organisation a way to maximise workforce engagement in putting sustainability initiatives into practice. And so, piece by piece, we put together a model of what a truly sustainable and sustaining organisation can be.

Cases can also show us how we can participate in giving shape to the future. Sustainability is as much about process as it is about destination. Cases help us answer the question of *how to* and provide positive prescriptions for action. They begin to map the path we can take to get to a world of sustainable organisations. Later in this chapter we provide one such model that we have constructed in this way and we show how a model of this kind can provide a meaningful context for learning from a case collection such as this.

## Who will find this book useful?

Several kinds of people will find this book useful. First, this is a book by researchers who were brave enough to venture out from the rarefied atmosphere of academia to enter into the everyday worlds of activities such as manufacturing and funeral services, banking and water supply, building construction and information technology. One of the uses of this book is therefore as an introduction for other researchers showing how sustainability research is done in the field; not field research among the indigenous inhabitants of remote islands of the Pacific or the jungle fastness of the Amazon but in what is for most people today, equally unknown territory – the interstices of the modern economy. There is knowledge to be gleaned here about gaining entry to a variety of organisations and how to undertake research in the workplace. There is information too on the theoretical perspectives that the case researchers have brought to provide focus in their work and how those perspectives influenced the questions asked and the data gathered.

This book will also be useful to teachers and students in tertiary level business courses, to the designers and teachers of in-house company executive programs and the participants in those programs. In the training of people for a profession, what brings a course alive is the use of contemporary examples of leading-edge practice. The cases needed for exciting learning experiences are not ideal examples but the partly successful, partly flawed human attempts at innovation in real life. The students can be challenged to find the difference between what has worked and what has not, to examine emergent models, to appreciate and critique them and to ask what would be needed to make this all work better next time. This becomes an adventure in analytic thought, in diagnosis, redesign and remediation; in applying theory to practice to improve practice and practice to theory to reformulate theory.

Where the participants themselves hold jobs, the cases can be a point of departure for exploring what is being done about sustainability in their own organisation, what is not being done, and how what has been learned from a case could be usefully applied in their own organisation at their initiative. Cases therefore can become a galvanising force, enlarging the cloud of sustainability change agents.

This book will also be useful to a range of professionals who are increasingly involved in implementing change toward more sustainable organisational practices. As we have pointed out above, sustainability has become a major talking point in conferences and seminars, both in business and in technology and science. There is a time to talk and a time to stop talking and act. In the words of an African proverb: "You don't grow pumpkins by just talking about them". We find an increasing impatience amongst consultants, managers and technologists with the continuing talk about sustainability. These cases provide examples of professionals of many kinds who have gone beyond talk to action – thoughtful, informed and often effective action. A range of professionals will be able to identify with this and find models here to emulate, inspiration to provide leadership in driving sustainability forward in the organisations in which they are involved. 'We can make it happen' is the clear message of many of these cases.

Such an approach is highly relevant in this United Nations Decade of Education for Sustainable Development (2005–2014) and our cases are useful tools with which educators, trainers and organisational development practitioners (OD practitioners) can implement the principles of education for sustainability. That is, they are designed "to help people to develop the attitudes, skills and knowledge to make informed decisions for the benefit of themselves and others, now and in the future, and to act upon these decisions" <http://www.unesco.org/en/esd/>. We hope they will build the capacity and prompt the awareness raising and critical thinking that will enable the workforce and citizens in general to adopt sustainable modes of production and consumption (Wals 2009).

## How to use this book

This book contains thirteen case studies. The index lists the case studies by title and also indicates the industry which provides the context for the case. There are, for example, two cases from the banking industry here – Westpac Banking Corporation and the Bendigo and Adelaide Bank. Those researchers or teachers with an industry focus such as this can immediately find the cases most relevant to their interests. In fact, these two cases provide a potentially very interesting comparison for someone interested in the finance sector as, while the two banks share a commitment to sustainability, they are pursuing two very different business strategies and so their approaches to sustainability also differ.

Other researchers, with a focus such as how to generate workforce engagement in sustainability, will need to read the cases thoroughly to find which are most relevant to their own theoretical focus. For example, there is something relevant to workforce engagement in most of the cases. To take just two cases: the Westfield case (Case 11) deals with the way in which a planned change process created a measured improvement in workforce engagement in the Finance division of that firm; and the Yarra Valley Water case (Case 13) shows how a systematic culture-building process in this government utility contributed to workforce engagement in sustainability initiatives and their implementation. In this way researchers can use the cases to find material for secondary analysis and theory building.

The cases are also relevant for teaching in business and professional courses. We have already referred to this above. It would be possible for example to base an entire introductory sustainability course on the study of just the cases in this book. There are more than enough cases here to provide for a case analysis each week of a normal university term. One advantage of the case method is that it lends itself readily to adaptation for use with a variety of theoretical perspectives and the same case can also be used for different topics. For example, Andre Taylor's *City of Mandurah: Champions of change* (Case 2) could be readily used in a learning unit on leadership or in a unit on water management. It could also be used in a management course to illustrate how sustainability can be introduced at the local government level or in a research methodology course to illustrate field methods in sustainability research. A number of the cases, such as the Hewlett-Packard, IKEA and Indian garment industry cases, could be used in supply chain, operations management and international business teaching units, or would be relevant to executive development courses for multinationals. This book is a repertoire of relevant course materials with few limitations for an imaginative educator.

For the professional who is actively involved in contributing to a particular organisation as a member of the organisation or as a consultant, each case is a potential source of insight into how to intervene in an ongoing organisational system to move the organisation further along the path to sustainability. There are plenty of examples of how initiatives can be taken, influence exercised and blocks and constraints can be experienced in the change process. There is inspiration here: 'If others have done it, so can I'. There are also warnings: 'It isn't always as easy as it seems'.

Perhaps the major lesson to be learned from these cases applies to all of us. It is that the greatest difficulty we face is to see with new eyes our own organisational world that we think we know so well – the world can look very different through a sustainability lens. Looking at our organisational practices with this new perspective can lead us to ask: 'Why have we always done it this way?' and 'Is there another way to do this that would be less destructive to the planet, to the community and which would contribute to a world fit for future generations to live in?' or even 'Should we be doing this at all?'. There is another challenging insight to be gleaned from these case studies: if we want others to change, we must first change ourselves. We must embody the change we want to bring about for what we do shouts louder than what we say.

# Summary of case studies

## Short case studies

### Bendigo Bank's approach to sustainability:
### Successful customers and successful communities create a successful bank

The Bendigo and Adelaide Bank (B&AB) is widely acknowledged for its customer and community engagement activities. This chapter describes the collaborative community engagement business strategy that has enabled B&AB to rapidly expand its business while addressing local community needs.

### Hewlett Packard's supply chain

The HP case outlines the innovative strategies adopted by Hewlett Packard in greening its supply chain. It focuses on HP as a global company and associated global supply chain issues and describes some of the issues in establishing new business models such as product service systems.

### Indian clothing industry: Ethical and social responsibility dilemmas

The Indian clothing industry case highlights the relationship between globalisation, offshoring of work and sustainability. The case shows how simply adopting ethical codes of conduct in the Indian clothing industry has failed to tackle the underlying problems such as child labour and why indeed such efforts can be counterproductive.

### Leighton Contractors: Becoming a sustainable organisation

The Leighton Contractors case study describes how one of Australia's largest engineering companies has begun the process of change for sustainability. The case study highlights the implementation of sustainability as a complex change program that requires flexibility and adaptability in the field; theoretically it illustrates processual theories and the role of ethics in change for sustainability.

### Westpac Banking Corporation: What do we mean by sustainability?

This case explores the challenges in gaining a shared understanding of corporate sustainability amongst employees, customers and other stakeholders and how this can impact organisational change towards sustainability. It raises questions around how can companies best communicate to customers and to employees on sustainability.

### Yarra Valley Water: Learning and change for sustainability

The Yarra Valley Water case highlights the need to link business strategy with sustainability tools and culture. It shows how these three aspects of a change management approach can mutually reinforce ongoing and effective learning and result in impressive business and sustainability outcomes.

# Long case studies

## City of Mandurah: Champions of change

*City of Mandurah: Champions of change* uses a case study from the Australian water industry to explore the role of change agents (also known as 'champions') who can be instrumental in triggering and driving change to advance more sustainable practices. Knowledge of the many individual, group-based and contextual factors that help such leaders to emerge and effectively exert influence has practical significance. It can be used to create supportive organisational contexts and customised interventions (e.g. leadership development programs) to build their capacity to initiate and drive change.

## Fuji Xerox Australia Eco-Manufacturing Centre: A case study in strategic sustainability

The Fuji Xerox case shows how technological advances, supportive leadership and a high performance workplace culture can lead to success in strategic sustainability. It tells a story that goes beyond remanufacturing towards the implementation of total product responsibility.

## IKEA: A company's progression to a strategic approach

The IKEA case has a number of interesting elements that contribute to the study of sustainability practice. This international home furnishing group was a pioneer in initiating policies that addressed issues of social and environmental sustainability practice. They were early in recognising the need to take affirmative actions in cooperation with other members of their supply chain. IKEA have now moved on to adopt a more integrated strategy of sustainability management.

## Interface's approach to sustainability: Manufacturing green carpet

This case discusses how Interface, considered to be a global leader in environmental sustainability, has integrated sustainability into its business. It describes three major, and inter-connected, components of Interface's strategy, which has resulted in significant business and sustainability outcomes.

## Ports of Auckland's response to climate change related challenges

The Ports of Auckland case study presents an illustrative case of company climate change response. It discusses environmental monitoring and collaboration options that might be considered as part of its future climate change strategy to address sustainability concerns.

## State of Grace: Can death be sustainable?

State of Grace is a case study of a start-up business that was beginning to gain momentum. The two women founders were trying to find the balance between how much and how fast to grow the business, whilst attempting to meet their family commitments and stay true to their sustainability values. An underlying issue in the case is their prioritisation of who and what had the greatest need for sustaining.

## Westfield talent management: Creating a high performance culture in the Australian Operations Finance division

This case documents the remarkable turnaround of the Australian Operations Finance division in Westfield Group Limited. The Australian Finance division was responsible for supporting the operational business, and in 2005 it became apparent that the function needed a significant revamp to achieve efficiencies and provide high level support for the strategic development of the company. The case describes how the organisation rebuilt the function over four years, upgraded the efficiency and quality of its operations and significantly improved performance.

# A context for the cases

At this point we wish to introduce the reader to a systematic model of the path to sustainability which we have built over some years. This model arose out of our involvement in many organisations aspiring to move toward more sustainable practices. We use the model here to provide a systematic context for placing the cases in relation to each other so we can identify their potential contribution to helping define a path to sustainability. We want to stress that this is only one of many possible ways to approach the cases. This is simply an illustration of how a theoretical model can extract value from apparently disparate cases, allowing us to maximise our learning from the variety of sustainability initiatives taken.

The model is a revised and updated version of that originally published in Dunphy, Griffiths and Benn (2007). The model, The Sustainability Phase Model, provides a set of six distinct steps which together represent a path that organisations take in progressing toward sustainability. In its simplest form, the model has six phases as shown in Figure 1.

**Figure 1** Overview: The Sustainability Phase Model

| **Phase 1** | **Rejection** | - | the freeloaders and stealthy saboteurs |
| **Phase 2** | **Non-responsiveness** | - | the 'bunker wombats' |
| **Phase 3** | **Compliance** | - | the reactive minimalists |
| **Phase 4** | **Efficiency** | - | the industrious stewards |
| **Phase 5** | **Strategic proactivity** | - | the proactive strategists |
| **Phase 6** | **The sustaining corporation** | - | the transformative futurists |

*Source*: *Organisational Change for Corporate Sustainability*, Routledge, London and New York, 2003, revised edition, 2007.

Since publishing the model in 2007, we have revised and expanded it on the basis of subsequent research. The basic model remains the same. However to clarify the main sources of performance enhancement that can be achieved at each phase, we have added specific kinds of value which can be added at each phase and specific types of waste to be targeted. We also give the central theme which defines the character of the phase (see Appendix A). The original model concentrated on the changes from stage to stage in both human and ecological sustainability. Human sustainability refers to practices which contribute to the personal wellbeing and up-skilling of the workforce and to contributions made by the organisation to the wellbeing of the community and society as a whole. Ecological sustainability refers to the positive impacts of the organisation's activities on the environment by, for example, eliminating emissions and recycling water. In the new version of the model, we have made the corresponding economic changes more explicit - a contribution made by Bruce Perrott (see Figure 2).

This version of the sustainability management model (Figure 1) identifies the three streams that are involved in creating sustainable organisations namely: human, economic and environmental. The model shows a starting point to the left where an organisation has existing attitudes, mores and values regarding sustainability. There are pressures to change these existing values that can be described as the drivers for change towards being a more sustainable organisation. These pressures or drivers may come from external stakeholders, industry or government sources in the form of their changing attitudes to sustainability. Drivers may also stem from internal sources, as management and staff themselves realise the need to change organisational standards and behaviour for sustainability management. Along the sustainability pathway, there may be input and consultation with key stakeholder groups.

Management may begin to conduct audits of sustainability practice, and initiate strategies to achieve new objectives in the three key streams of sustainable management. The model shows six possible stages of progression along the path to becoming a sustaining organisation. These six stages are described in detail below. Along this path there will be ongoing drivers that dictate the nature and momentum for further change. There will also be barriers that inhibit or block the direction and rate of change. At the right hand end of the model there is a cloud which is intended to represent the evolving and changing vision of what their sustainable organisation may be working towards. This vision must necessarily change and evolve as standards and expectations change, in addition to the resources and capabilities that make the realisation of that vision possible.

Any particular organisation may progress along the three sustainability streams at different rates. For example, an organisation may be progressive and successful in moving to a fairly advanced phase (e.g. Phase 4 or 5) of environmental sustainability but lag in the other two streams. We emphasise that this is an ideal model and that in the complex organisations of today we do not expect to see a linear progression towards sustainability. There may be trade-offs between the different dimensions of sustainability (Angus-Leppan *et al.* 2010); organisations may leapfrog or retreat across phases, or their different divisions may move at different rates. But in our experience, the model acts as an extremely useful heuristic – a tool for understanding different practices for organisational sustainability.

We now summarise the key features of each of the phases and the reader is referred to Figure 1 for the position of the phase in the overall phase model and to Appendix A for more detail. In particular, Appendix A gives more details of the kinds of action that are typical for the phase. We draw on and supplement the original phase descriptions from Dunphy *et al.* (2007) as published in *Organizational Change for Corporate Sustainability*.

**Figure 2** Strategic Sustainability Phase Stages

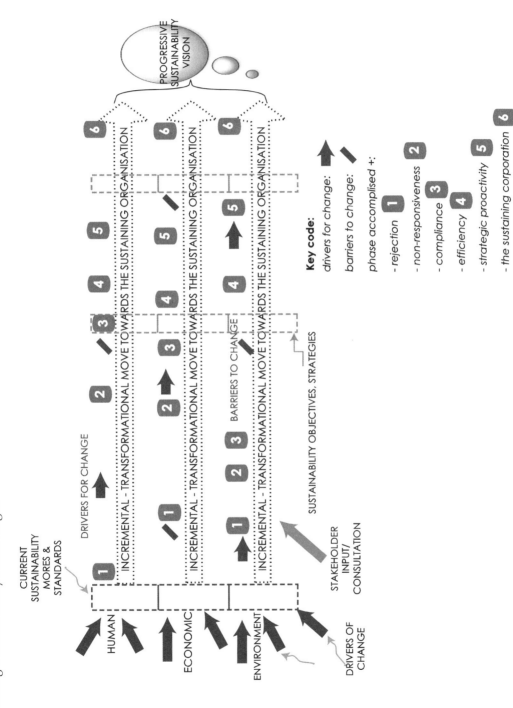

So what are the distinguishing characteristics of each of the phases?

*Phase 1 Rejection:* The senior managers in these organisations take the attitude that ecological and social resources exist to be exploited by the organisation for immediate economic gain. As far as human sustainability is concerned, employees are regarded as 'industrial cannon fodder'; there is no commitment to anything more than the minimal skill development needed to get the work done, and occupational health and safety measures are ignored or given 'lip service'. On the ecological side, managers disregard destructive environmental impacts of the organisation's activities and expect the community to pay the costs of any remediation required. There is a strong belief that the firm exists to make profit and other claims by the community are rejected as illegitimate. Spokespersons often actively reject the idea that climate change is occurring or if they accept that it is, argue that it is not caused by human activity. Managers strongly resist any attempts by governments or community activist groups to place constraints on the organisation's activities.

In terms of the economic dimension, these organisations have a short-term perspective. They operate a business model that is generating income from what was a timely product or market opportunity. There is little thought about the medium-term future of the organisation or what should constitute a sustainable business model. There is an operational focus on the day-to-day aspects of the organisation and no effective strategic planning around the ongoing growth and development of the business.

We refer to these organisations as 'stealthy saboteurs and freeloaders' because their opposition sabotages movement toward a more sustainable world. The costs of modifying prevailing practices are passed on to other organisations and to the community.

The prevailing theme for this Phase 1 is: *exploit resources for maximum short-term gain.*

*Phase 2 Non-responsiveness*: Organisations in this phase are usually there because of the lack of awareness or ignorance of senior executives about sustainability, rather than from their active opposition to adopting a corporate ethic broader than financial gain. Many of the organisations in this category embody the culture of the past century, concentrating on 'business as usual', operating in conventional ways that do not incorporate sustainability into decision making. Human resource strategies, if they exist, are focused mainly on creating and maintaining a compliant workforce. Where possible, community issues are ignored and any negative impacts on the environment of the organisation's activities are disregarded. Thus many of the true costs of the organisation's activities are externalised.

In terms of the economic dimension, these organisations may be aware that they need to create a sustainable business model for the longer term viability of the enterprise if only to satisfy the longer term expectations of key stakeholders. This awareness is often triggered by the realisation that the original growth momentum experienced by the organisation may have a limited time scale. However, no time or expertise has been made available to action this awareness and convert it into sustainable strategic ideas for the future of the organisation. So the culture is still focused on short-term operations and results.

We refer to these organisations as the 'bunker wombats' as, like wombats, they prefer to avoid what is obvious in the light of day and hunker down in dark bunkers away from where the action is taking place.

The prevailing theme of this Phase 2 is: *business as usual.*

Organisations in Phases 1 and 2 are essentially relics of the past and increasingly experience costs of non-compliance including serious targeting by environmental and social activists, fines, increasing costs of capital and increasing difficulty in recruiting and retaining first class employees. In this book, we present cases of organisations that have largely progressed to Phase 3 and beyond. While in that way they are success stories, the cases each illustrate the challenges that particular sectors face in attaining sustainability while remaining competitive in the short term. The Indian Garment Industry case, for example, highlights some of the challenges faced by textile, clothing and footwear sector organisations.

*Phase 3 Compliance:* This phase focuses on reducing the risk of sanctions for failing to meet the minimum standards as an employer or producer. An organisation in this phase is seeking to be compliant to the law in relation to environmental, health and safety regulations and relevant and legitimate community expectations. Senior managers focus on being 'a decent employer and corporate citizen' by ensuring a safe, healthy workplace and avoiding environmental abuses that could lead to litigation or strong community action directed towards the firm. There is usually little integration between HR and environmental functions. The organisation may support community charitable ventures. However the organisation's policies are primarily reactive to growing legal requirements and community expectations for more sustainable practices rather than focused on taking voluntary initiatives. If approached by governments to participate voluntarily in government sponsored sustainability initiatives, they may however collaborate.

In terms of the economic dimension, these organisations have made a start on the planning that will be necessary to create continuity and build the ongoing viability of the organisation. Planning is usually carried out as a matter of compliance with what is expected of an organisation of their particular position and status. These plans may also be required by key stakeholder groups such as banks. However, planning and strategy for the medium to long term tends to be ad hoc and lack continuity and the ongoing commitment of the senior managers. Apart from the risk minimisation aspect, other forms of value added in this phase result from the positive move to become more sustainable, particularly by providing easier access to finance, improved relationships with regulators and the basis for a positive reputation as a worthy corporate citizen.

We refer to these organisations as 'reactive minimalists' as they accept the demands of the environment to move toward more sustainable practices but limit their responses to what is required.

The central theme of Phase 3 is: *avoid risk.*

*Phase 4 Efficiency:* This phase reflects a growing awareness on the part of managers that there are real advantages to be gained by proactively instituting sustainability practices. The focus is on progressively eliminating waste and increasing process and materials efficiencies. In particular, human resource and environmental policies and practices are used to reduce costs and to increase efficiency. There may be: systematic investment in training and development to build a more skilled workforce, which is seen as a means to higher productivity and efficiency; introduction of new technologies to increase the efficiency of resource use and reduce emissions, water consumption, energy consumption etc.; redesign of production and distribution processes to minimise delays and avoid duplication of activities; product and packaging redesign to progressively reduce and eliminate waste; and, improvements in product and service quality to add value. While moves towards sustainability may involve additional expense, they can also have significant payoffs in terms of reducing unnecessary costs and generating new income directly or indirectly. This is the beginning of making sustainability an integral part of the business. We refer to these organisations as 'industrious stewards' as they are primarily concerned with ensuring that waste is eliminated and resources used effectively.

In terms of the economic dimension, these organisations have produced an effective business model that is potentially capable of sustaining the organisation's operation into the future. They have given thought to risk management approaches that will ensure the continuity and ongoing viability of operations. This may include some early planning for knowledge management and succession management. However, there is still a focus on operational continuity rather than the elevated strategic framework that will be necessary for longer term ongoing viability and survival in a rapidly changing world. These organisations begin to experience some of the economic benefits resulting from the numerous social and environmental sustainability initiatives taken. The main kinds of value added come through cost reduction and savings, increased employee productivity, greater employee involvement and engagement and better teamwork and lateral communication. The savings come from conservation of physical resources such as water, energy, heat, power and materials and from a reduction in human resource 'waste' such as underutilised people, costly staff turnover, absenteeism, lack of motivation and commitment, internal conflict and the politicisation of the workplace. In this book, the Leighton Contractors, IKEA and Ports of Auckland cases illustrate many of the principles of the Efficiency phase.

The central theme of Phase 4, the efficiency stage, is: *do more with less.*

*Phase 5 Strategic proactivity:* This phase moves the organisation further along the sustainability path by making sustainability central to the enterprise business strategy. Senior executives begin to see sustainability as providing a potential competitive advantage and try to position the organisation as a leader in sustainable business practices including good corporate governance and corporate citizenship. The focus is on innovation. Advanced human resource strategies are designed to make the firm an 'employer of choice' so it attracts high level talent. Corporate citizenship initiatives are used to build stakeholder support, and innovative, quality products are introduced that are environmentally safe and healthy. The commitment to sustainability is, however, strongly embedded in the quest for maximising longer term corporate profitability, that is, it is motivated primarily by intelligent corporate self-interest.

In terms of the economic dimension, these organisations have a strong capability to plan and achieve ongoing financial stability. They have made effective strategic planning and management an integral part of their modus operandi. However, because they lack an overall philosophy of sustainability, they may move from one major project or business opportunity to the next without these initiatives necessarily having continuity. (This distinguishes them from the sustaining corporation described in the next phase). This project focus may result in some discontinuity of performance over the medium to longer term. At this phase organisations incorporate the general concept of sustainable corporate responsibility into their ongoing strategic planning and management. They also see the strategic need to include supply chain partners and to collaborate with organisations from the not-for-profit sector in their sustainability strategies. There is a realisation of the strategic benefits of embracing ongoing sustainable practice.

The main value added at this phase comes through increased revenue and market share, stronger stakeholder support, higher retention rates of competent staff, faster attraction of new customers, an established lead in developing new markets, becoming an employer of choice so attracting and retaining skilled managers and professionals, and operating at the high value-added end of the market. Conversely there can be: increased targeting of the waste of losing market share and revenue streams (often associated with being a laggard rather than a leader); reduction in waste of executive time that occurs with having to deal with hostile or competitive stakeholders; and, avoidance of costs associated with maintaining increasingly redundant and unprofitable operations and units embedded in the old world, and of operating at the low value-added end of the market.

We refer to these organisations as 'proactive strategists' because they see sustainability as integral to business strategy and actively pursue its business advantages. Examples of cases from the book whose management practices typify this phase include Fuji Xerox, Westpac and Hewlett-Packard.

The central theme of this Phase 5 is: *lead in value-adding and innovation*.

*Phase 6 The sustaining corporation:* In this final phase, senior executives and the majority of the members of the organisation have strongly internalised the ideology of working for a sustainable world. As each advance in this area supports the developing strategies of the organisation, they are redefining the business environment. If the organisation is a for-profit company, it still pursues the traditional business objective of providing an excellent return to investors but voluntarily goes beyond this by actively promoting ecological sustainability values and practices in the industry and society generally. The nature of the corporation is reinterpreted as an integral self-renewing element of the whole society in its ecological context. Its fundamental commitment is to facilitate the emergence of a society that supports the ecological viability of the planet and its biodiversity and also contributes to just, equitable social practices and human fulfilment.

These organisations have developed the capability to create a business model that provides ongoing and continuing financial viability. Such organisations have generally diversified to an extent that ensures continuity of performance for the whole organisation. There is ongoing and integrated knowledge capture, storage and dissemination of the ways the organisation sustains growth and viability. Continuity of performance is achieved by the means of reliable and diverse sources of finance and human capital. Stakeholder involvement is ongoing and engagement is a strong and accepted aspect of the culture. There is an integrated approach to coordinate strategies in the three main streams of sustainability: economic, social and environmental. All key members of the supply chain are involved in well-coordinated sustainability practices including a focused effort to improve the sustainable behaviour of customers and consumers. To achieve continuity of sustainable performance, effective change management becomes an ongoing and effective 'built-in' capability.

The potential business benefits of this phase are that the organisation is seen as exercising leadership for the global sustainability movement; its reputation is enhanced and it continues to build reputation and stakeholder support and involvement. If it is a for-profit organisation, its share value increases and it increasingly attracts and retains the most talented and highly motivated employees available.

We refer to these organisations as 'transforming futurists' because they are not only concerned with the ongoing transformation of their own organisations to align with the requirements of a more sustainable world, but they are also actively involved in transforming the larger economy and society in the same direction. A number of our cases are examples of sustaining organisations. They include Interface, Bendigo Bank and State of Grace.

The central theme of this Phase 6 is: *Transform ourselves: lead in creating a sustainable world.*

This fuller version of the model which we have described above includes: a clear statement of the overall objective for each phase; the key business opportunity; typical actions that are taken by organisations to pursue the objective and realise the opportunity; and, finally, the value that can be added for the organisation and potential sources of savings by the elimination of various kinds of waste.

## Conclusion

The cases presented here are diverse – representing different positions not only on our phase model but different sectoral and other constraints and theoretical approaches to understanding such challenges. While the authors of some of the cases have not presented a particular theoretical framework, cases of this kind can also be very useful in teaching because students can be encouraged to bring differing theoretical frameworks to bear on the process of interpreting the information provided about the organisation's sustainability activities. In our opinion a casebook of this kind should provide for diverse approaches to teaching sustainability rather than imposing one particular viewpoint. At the same time, we need to strive to develop an informed and emerging consensus, based on empirical examples such as these cases, on the most effective way that organisations can advance toward supporting the life systems of the planet and a just and equitable society living and working in harmony with processes of regeneration in the natural world.

So welcome to this collection of cases. We hope that you find reading the cases stimulating and informative and that you are led to undertake further case research yourself, and to make cases an important aid to your teaching, being inspired to become a change champion in creating more sustainable and sustaining organisations. The world needs more visionary challengers and change masters and more effective practical contributions to creating a sustainable future. We recognise that much more specific detail is needed of the ways that change managers and leaders can overcome sustainability barriers such as the short-termism of some stakeholders and arguments about the meaning of sustainability. Our future publications will address these avenues for research and teaching.

# Appendix A: Details of phases 3–6

## Phase 3: Compliance - the reactive minimalists

**Objective**: Seek to be compliant to the law and all environmental, health and safety requirements and relevant community expectations.

**Business opportunities**: Avoid the potentially huge costs of non-compliance and create an effective risk management system.

**Typical actions**:

- determine what is relevant legislation, regulations and community expectations
- build an effective risk management system with an informed workforce committed to compliance
- establish an organised measurement and monitoring system.

## Phase 3: Compliance - the reactive minimalists

**Value added**:

- risk minimisation
- easier finance
- basis for positive reputation
- improved relationships with regulators.

**Waste to target at this phase**:

- fines for non-compliance
- higher-cost finance
- poor reputation
- time and energy wasted coping with antagonistic regulators and community groups.

## Phase 4: Efficiency - the industrious stewards

**Objective**: Progressively eliminate waste and increase process and material efficiencies.

**Key business opportunity**: Increase efficiencies by waste reduction and reorganisation.

**Typical actions**:

- reduce resource use (energy, water, materials)
- design/redesign buildings/plant to dramatically reduce 'footprint', create adaptable spaces
- move to front-of-pipe solutions to eliminate waste or return it to the production cycle as a resource (biomimicry)
- recycle/remanufacture (life cycle stewardship; cleaner production)
- dematerialise – service provision rather than material production
- redesign products: sustainably produced and environmentally friendly
- meet international Global Reporting Initiative (GRI) guidelines.

## Phase 4: Efficiency - the industrious stewards

**Value added**:

- cost reduction; saving
- increased employee productivity
- increased employee involvement/engagement
- better teamwork and lateral communication.

**Waste to target at this phase**:

- wasted physical resources, e.g. water, energy, heat, power, materials
- wasted human resources, e.g. under-utilised people, turnover of important skills, absenteeism, lack of motivation, engagement, commitment, internal conflict and political processes.

## Phase 5: Strategic proactivity - the proactive strategists

**Objective**: Pursue the strategic opportunities in sustainability.

**Key business opportunity**: Become market leader through pursuing the strategic potential of sustainability.

**Typical actions**:

- commit strongly to sustainability
- re-brand and build wider stakeholder support
- be early in on new product/service demand curves
- creatively destroy existing product designs, manufacturing models and re-invent the organisation, leapfrog competition by early breakthroughs
- increase employee and stakeholder engagement to source innovation ideas
- shift the prevailing business paradigm in environmental and social ideas
- innovate with new models of stakeholder governance
- concentrate on adding value and innovating.

## Phase 5: Strategic proactivity - the proactive strategists

**Value added**:

- increased revenue and market share
- stronger stakeholder support (reputation and commitment)
- higher customer retention rates; faster attraction of new customers
- established lead in developing new markets
- employer of choice – attract and retain skilled managers and professionals
- operate at high value-added end of market.

**Waste to target at this phase**:

- lost revenue and market share
- hostile or apathetic stakeholders
- loss of customers
- failure to enter and secure a place in new markets
- low skilled managers and employees
- operations at the low value-added end of the market
- redundant operations and units embedded in the old world.

## Phase 6: The sustaining corporation - the transformative futurists

**Objective**: Redefine the business environment in the interests of a more sustainable world and to support core strategies of the organisation.

**Key business opportunity**: Create a constructive culture that continually renews the long-term viability of the organisation.

**Typical actions**:

- participate in changing the 'rules of the game' to achieve sustainability
- participate in the public policy formation
- reorganise the company's supply chain to ensure that the whole production process is sustainable
- build human and relational capital
- support dematerialisation and the growth of the knowledge-based economy
- model best practice; support/publicise best practice elsewhere
- participate in international agreements
- seek external auditing of sustainability
- influence capital markets to support long-term value-adding
- build a constructive culture that encourages openness, debate, innovation and participation.

## Phase 6: The sustaining corporation - the transformative futurists

**Potential business benefits**:

- global leadership of the sustainability movement
- enhanced reputation and stakeholder support and involvement
- increased share value
- attraction/retention of talented, highly motivated employees.

**Waste to target at this phase**:

- strategic diversion from the sustainability goal for the organisation and society
- products, services, processes that damage reputation as a sustainability leader
- loss of business focus
- non-alignment of corporate talent with the organisation's strategic gaols; loss of critically important talent
- any remaining non sustainable work processes, products or services.

# References

Angus-Leppan, T, Benn, S & Young, L 2010, 'A sensemaking approach to tradeoffs and synergies between human and ecological elements of corporate sustainability', *Business Strategy and the Environment* 19 (3): 230-244.

Doppelt, B 2009, *Leading Change toward Sustainability: A change management guide for business, government and civil society*, Greenleaf, Sheffield UK, rev. edn.

Dunphy, D, Griffiths, A & Benn, S 2007, *Organizational Change for Corporate Sustainability: A guide for leaders and change agents of the future*, Routledge, London and New York, 2nd edition.

Flannery, T 2002, *The Future Eaters*, Grove Press, NY.

Flannery, T 2009, *Now or Never: A sustainable future for Australia*, Black Inc, Melbourne.

Flannery, T 2010, *Here on Earth: An argument for hope*, Text Publishing, Melbourne.

Friedman, T 2009, *Hot, Flat, and Crowded*, Penguin, London.

Galea, C (ed), 2009, *Consulting for Business Sustainability*, Greenleaf Publishing Ltd, Sheffield.

Hamilton, C 2010, *Requiem for a Species: Why we resist the truth about climate change*, Allen & Unwin, Sydney.

IBM Corporation 2008, 'The Enterprise of the Future', Global CEO Survey, IBM Global Business Services, NY.

IBM Corporation 2008, *Making Change Work*, IBM Global Services, NY.

IBM Corporation 2010, *Capitalising on Complexity*, Global CEO Survey, IBM global Business Services, NY.

Jones, G, Dunphy, D, Fishman, R, Larne, M & Canter, C 2006, rev. edn. 2011, *In Great Company: Unlocking the secrets of cultural transformation*, Human Synergistics, Sydney (revised edition in press).

Kahane, A 2010, *Power and Love: A Theory and Practice of Social Change*, Berrett-Koehler, San Francisco.

Lovelock, J 2009, *The Vanishing Face of Gaia: A final warning*, Allen Lane, London.

Monbiot, G 2006, *Heat: How we can stop the planet burning*, Penguin, London.

O'Brien, J (ed), 2009, *Opportunities Beyond Carbon: Looking forward to a sustainable world*, Melbourne University Press, Melbourne.

Porter, T 2008, 'Managerial applications of corporate social responsibility and systems thinking for achieving sustainability outcomes', *Systems Research and Behavioural Science* 25(3) 397-411.

Porter, T & Cordoba, J 2009, 'Three Views of Systems Theories and their Implications for Sustainability Education', *Journal of Management Education*, 33(3) 323-347.

Spratt, D & Sutton, P 2008, *Climate Code Red: The case for emergency action*, Scribe, Melbourne.

<http://www.unesco.org/en/esd/>, viewed 15 January 2011.

<http://www.unglobalcompact.org/news/42-06-22-2010>, viewed 3 January 2011.

Wals, A 2009, 'Review of Contexts and Structures for Education for Sustainable Development', UNESCO, France.

# About the Authors

## Suzanne Benn

Suzanne Benn was appointed in 2011 to the newly created Chair in Sustainable Enterprise at the Business School of the University of Technology, Sydney. As Professor of Sustainable Enterprise, she will provide a focus for sustainability research, teaching and external engagement across the faculty and with other parts of the university. The role will be a springboard for collaboration with external stakeholders in business, government and NGOs.

Suzanne was previously Head of the Graduate School of the Environment, Professor of Education for Sustainability and Director of ARIES at Macquarie University, Sydney. Suzanne first studied Chemistry and Biochemistry at the University of Sydney, followed by her postgraduate studies at the University of New South Wales, gaining her PhD there in Environmental Studies. She has wide experience across industry and the range of educational sectors, having worked in the chemicals industry, and taught in the school, technical and tertiary sectors. Before coming to Macquarie, she was Environmental Education Coordinator at the University of New South Wales and more recently, spent ten years researching and lecturing in the area of corporate social responsibility, corporate sustainability and organisational change in the School of Management at the University of Technology, Sydney.

## Dexter Dunphy

Dexter Dunphy is Emeritus Professor and Senior Associate in the Centre for Corporate Governance, Faculty of Business, the University of Technology, Sydney. Throughout his career, Dexter has researched and consulted in the areas of the design and implementation of corporate change, human resource management and corporate sustainability. His research in these areas is published in over 90 articles and 24 books. He has held professorial positions at various universities in Australia and overseas and continues to combine his academic interests with practical involvement in business, government and not-for-profit organisations. He works actively with executives and managers to initiate more sustainable and sustaining practices in their organisations and to ensure that these innovations are then more widely adopted.

## Bruce Perrott

Bruce Perrott is Senior Lecturer in the Faculty of Business at the University of Technology, Sydney. He is interested in how business and marketing strategies are formulated, formalised and managed through to the implementation phase. This interest is pursued through a broad range of activities including: advice and consultation to senior management on strategic direction, methodology and process; research projects that investigate sustainable strategic process, management and outcomes; education and management development on all aspects of strategic management and strategic marketing in undergraduate, postgraduate and management forums. His PhD research was a study of strategic management processes in state owned enterprises.

# About the Contributors

## Tamsin Angus-Leppan

Tamsin Angus-Leppan is a Project Leader at the Australian Research Institute for Environment and Sustainability (ARIES), Macquarie University, Sydney. She earned her PhD in 2009 on the topic of stakeholder sensemaking and CSR. Her areas of research interest include CSR, sensemaking, ethical consumerism, sustainable development, cross-cultural research and qualitative methods. Her projects with ARIES have included the development of a sustainable business toolkit for SMEs in conjunction with the St James Ethics Centre and work with the Ethnic Communities Council of Australia to assess and develop methods for improving sustainable living in Sydney's ethnic communities. Her work has previously been published in the Journal of Business Ethics and Business Strategy and the Environment.

## John Chelliah

John Chelliah is a lecturer in strategy and entrepreneurship at the University of Technology, Sydney. He has a PhD from RMIT University, Melbourne. Prior to joining academia, he held senior management positions in several organisations located in the UK, USA, Malaysia, New Zealand and Australia. He has published widely in reputable international management journals covering strategy, management consulting, human resource management, change management, international management, supply chain management and entrepreneurship. John also contributes to management education through delivering lectures as an invited speaker at business schools in Europe, Asia and the South Pacific.

## Stephen Chen

Stephen Chen is an Associate Professor of Business at Macquarie University, Sydney. He obtained his MBA from Cranfield School of Management and his PhD in Management from Imperial College, London. He has previously taught at the City University Business School (now Cass Business School), Manchester Business School, Henley Management College, Open University (UK), UCLA and the Australian National University. His research interests are in the general areas of strategic management and international business, including corporate social responsibility in multinational firms, social entrepreneurship and management of not-for-profit organisations.

## Eva Collins

Eva Collins is an Associate Professor in the Strategy and Human Resource Management Department at the University of Waikato, New Zealand. She earned her PhD in Public Policy from George Washington University. Her teaching and research have focused on the sustainability practices of business, business climate change strategies and voluntary environmental programs.

## Patrick Crittenden

Patrick Crittenden is the Director of Sustainable Business Pty Ltd and a Project Leader at the Australian Research Institute for Environment and Sustainability (ARIES), Macquarie University, Sydney. Patrick has a background in business strategy, policy development and organisational change that he has applied over the past 15 years to the issues of energy efficiency, climate change and corporate sustainability. He combines detailed knowledge of current and emerging government policy and legislation with facilitation skills to build organisational capability and effective corporate and operational response to emerging issues such as climate change.

## Katrin Herdering

Katrin Herdering is a postgraduate student at Auckland University of Technology (AUT), New Zealand. In 2009, she was the top overall graduate in the Master of Professional Business Studies at AUT. She previously worked in Germany for almost eight years. In 2006, Katrin completed her job-related Bachelor of Arts in Business Administration at the University of Applied Sciences in Münster.

## Kate Kearins

Kate Kearins is Professor of Management and Associate Dean Research at Auckland University of Technology, New Zealand. She previously worked at the University of Waikato where she began her research into issues surrounding business and sustainability, environmental management, accountability and stakeholder engagement. This research continues to develop and has seen her win several international awards for case research, along with co-authors from AUT University and the University of Waikato.

## Robert Perey

Robert Perey is a Program Manager at the Australian Research Institute for Environment and Sustainability (ARIES), Macquarie University, Sydney. He has a background in organisational change and development with over 30 years' experience in senior management and consulting. Recent projects have ranged from biodiversity awareness in Culturally and Linguistically Diverse Communities (CALD), sustainability case study development for inclusion in MBA programs, to assisting a large not-for-profit organisation design and integrate sustainability into their strategy and operations. Robert's current research is investigating the organisational experiences of sustainability.

## Tony Stapledon

Tony Stapledon is an economist, architect and change leader with over 30 years' experience in senior management and consulting. In his consulting role as Group Sustainability Manager at Leighton Contractors he developed and implemented strategies to embed sustainability as a vehicle for continuity and growth, business improvement and corporate responsibility. He was formerly a research director at the Institute for Sustainable Futures at the University of Technology, Sydney, and from 1984 - 2001 was a Director of global design company Woods Bagot. His PhD in economics researched high performing sustainable organisations.

## Wendy Stubbs

Wendy Stubbs is a Senior Lecturer in the School of Geography and Environmental Science at Monash University, Melbourne. Wendy's research interests include corporate social responsibility, corporate sustainability, sustainable business models, education for sustainability and systems sustainability. Her research explores new business models that are grounded in the principles of sustainability (environmental, social and economic). She is the coordinator for the Master of Corporate Environmental and Sustainability Management program and teaches corporate sustainability in the Monash MBA program. Wendy has an MBA from the Wharton Business School and a PhD from Monash University.

## André Taylor

André Taylor is an environmental and social scientist. His PhD research examined particular types of leaders ('champions' or 'change agents') who played key roles in promoting more sustainable practices in the Australian water industry. Over the last 20 years, André has worked in five Australian states for consultancy firms, state government agencies, local government authorities and universities. He currently runs his own consulting business and is fortunate enough to work with enthusiastic environmental leaders around Australia to help them drive change.

## Helen Tregidga

Helen Tregidga is a Senior Lecturer in the Accounting Department at Auckland University of Technology, New Zealand. Her PhD thesis, completed in 2007, analysed the discourse of sustainable development within organisational reports. Her research interests include organisations and the natural environment, sustainable development, 'sustainability' reporting and discourse theory and method.

## Tim Williams

Tim Williams is Head of Strategy, Group Sustainability, the Westpac Banking Corporation. Tim has driven the Group's sustainability agenda and performance for well over a decade through advice and management across most areas of the sustainability agenda. Tim's prior professional background is corporate and public affairs. He has post-graduate qualifications in public affairs and management consulting. He is researching the nature of organisational change for sustainability and the role of leadership within that process.

# Case 1

# Bendigo Bank's approach to sustainability: Successful customers and successful communities create a successful bank

WENDY STUBBS

## Introduction

This case study describes Bendigo and Adelaide Bank's (B&AB's) social sustainability approach – the Community Engagement Model (CEM). The chapter focuses on three aspects of B&AB's strategy that contribute to the success of the CEM:

1.  keep capital local

2.  collaboration and cooperation

3.  values, leadership and culture.

Bendigo Building Society was founded in 1858 and converted to a bank in 1995. Bendigo Bank merged with Adelaide Bank in 2007 to form The Bendigo and Adelaide Bank Group (B&AB), Australia's seventh largest bank. B&AB is based in the central Victorian city of Bendigo, 150 kilometres north-west of Melbourne. It is a publicly listed company with a market capitalisation of around AU$3.3 billion, which ranks it among the top 100 Australian companies on the Australian Stock Exchange. It has a range of diverse businesses, including:

*   a retailer of banking and wealth management services to households and small to medium businesses

*   a wholesaler of mortgages to third party lenders

*   a large margin lending business (leveraged equities)

*   subsidiaries including Rural Bank (agribusiness bank) and Community Sector Banking (banker to the not-for-profit sector).

The bank holds assets under management of more than AU$47 billion, over AU$28 billion in retail deposits and has had 149 years of consecutive profits. B&AB has more than 460 branches, 1.4 million customers and 82,000 shareholders.

During the 1990s when the 'big four'[1] Australian banks closed one-third of their bank branches, Bendigo Bank was undergoing a strategic review in response to an increasingly concentrated and competitive market. It sought to differentiate itself and to find a unique value proposition for customers and communities. It concluded that for it to thrive in its market – primarily regional banking – it needed to contribute to building more sustainable communities. It developed a community-development business model, the Community Engagement Model (CEM) that focuses on building successful communities. Community Bank is the first business based on the CEM; the first community bank branch was launched in 1998. B&AB has now applied this model to address other community needs such as telecommunications, community sector banking, youth development programs, corporate giving and philanthropy.

Under its Bendigo Bank retail banking brand, B&AB's business strategy is focused on building strong communities – both financially and socially. In 2002, B&AB acknowledged that there was a third element vital to long-term community wellbeing – the environment. It launched Generation Green in 2004, which encompasses a suite of initiatives to help reduce B&AB's and its customers' and communities' environmental impacts (see Table 1.1).

**Table 1.1** Bendigo and Adelaide Bank's environmental initiatives

| Initiative | Description |
|---|---|
| Generation Green home and personal loans | Offer discounted interest rates for environmentally friendly homes and products. |
| Community Energy Australia | Applies the CEM to pool community demand for energy. The income can be invested in local development. |
| Community Fuel | The North Central Victorian Community Fuel Project plans to develop a locally-owned bio-diesel plant, using locally-grown oil seed. |
| Solar cities | B&AB is a member of the Central Victorian Greenhouse Alliance, a consortium working with 13 municipalities to trial a range of solar energy options. |
| Ban The Bulb (BTB) | Campaign facilitated by B&AB to mobilise community groups to replace light bulbs with energy-efficient bulbs. In 2009/10, communities replaced about 200,000 bulbs and received about AU$400,000. The company that provided the bulbs received carbon credits they could then sell to companies wishing to offset their carbon emissions. BTB is anticipated to save the communities about AU$15 million over a five-year period in reduced electricity costs. |

*Source*: Interviews and <www.bendigoadelaide.com.au>.

---

[1] ANZ, Commonwealth Bank, NAB and Westpac.

B&AB is widely acknowledged for its sustainability initiatives and its customer and community engagement activities. Bendigo Bank was recognised as the Most Sustainable Company in Australia in 2001 and 2002 by Ethical Investor Magazine and received a merit award in 2003 for Outstanding Achievement in Social Development. B&AB was named one of the world's top 20 sustainable business stocks by SustainableBusiness.com in 2009.

# Research methodology

Data were collected in the period 2004–2005 and in 2010 from in-depth interviews with B&AB staff, company briefings and other secondary sources. Fourteen one–two hour interviews were conducted in 2004–2005 and three forty-five minute interviews in 2010 with the CEO and staff from retail, information technology, strategy, community and solutions, banking and wealth, marketing and corporate affairs, finance, and community bank chairpersons. Notes were taken from annual general meetings in 2004 and 2005, and a presentation in April 2010. Secondary data were sourced from annual reports, quarterly earnings announcements, internal company documents, personal communications and the website. All interviews were recorded and transcribed and then coded to extract themes using qualitative data analysis methods (Strauss & Corbin 1998).

# The Community Engagement Model

Community Bank is a tangible example of the B&AB's community engagement and strengthening strategy. A community bank branch (CBB) is owned by the local community. Once the community has shown sufficient interest, a community steering committee is formed and a public company with limited shares is established. Shares are issued to raise funds from the local population to cover the setup costs and initial running costs of a CBB, about AU$750,000. Each branch operates as a franchise of B&AB, using the name, logo and system of operations of B&AB. The community company secures the branch premises, purchases fittings and systems, and covers the branch running costs such as wages, power and telecommunications. B&AB provides the banking license, the bank brand, training of staff, a core range of products and services, systems, marketing support, and administrative support. The CBB and B&AB share the revenue from the products and services sold through the community bank. The margin varies across the products and services depending on the amount of work done by each party. Generally, the CBB receives about 50% of the revenue from consumer products but less on business products (Stubbs & Cocklin 2007).

Twenty per cent of a community bank's profit can be distributed as dividends to its shareholders once the community bank has accumulated a net profit. Only local community members can be shareholders in the community bank. Eighty per cent of the profit is set aside to fund further community development projects. This 80/20 split is built into the community bank franchise agreement. One employee explained that the reason for this split was "to ensure there is a balance between being cooperatively spirited and commercially based". The community bank board of directors decides how to distribute the profits and which community projects to fund. The outcomes from the community bank model are summarised in Table 1.2.

**Table 1.2** Community banking results 2010 (AU$)

| Metric | Result |
|---|---|
| Number of community bank branches (450 bank branches in total) | 263 |
| Local capital raised | $124 million |
| Amount of revenue paid to community banks each year | > $100 million |
| Amount paid to community bank shareholders in dividends | $12 million |
| Amount contributed to community projects. | $40 million |
| Number of community bank customers | 410,000 |
| Number of local jobs created | 1,500 |
| Number of community bank shareholders | 61,000 |
| Amount community banks directly inject into their communities each year (through salaries and other local expenditure) | $60 million |

*Source*: Personal communication and <www.bendigoadelaide.com.au>.

In 2010, B&AB initiated a project to better understand and measure the economic and social impacts of the CEM, particularly the flow-on effects of the community investment.

## Keep capital local

B&AB identified 'capital drain' from regional areas as a barrier to sustaining prosperous, self-sufficient communities. Initially the motivation to open a CBB was because all the other banks had closed the branches in a town or suburb. A loss of financial capital can lead to a loss of social capital – "features of social organisation, such as trust, norms, and networks, that can improve the efficiency of society by facilitating coordinated actions" (Putnam, Leonardi & Nanetti 1993, p. 167) – as one executive explained:

> The capital markets concentration is extreme in this country… so the consequence of that capital concentration got us thinking. There was a bigger issue at a community level. Losing control of financial capital was one element but financial capital, intellectual capital, productive capital and human capital, totals up social capital.

The motivation now to open a CBB is to set up a community enterprise to deliver an income stream for community building activities, so that the community can address the 'big issues' and control the outcomes. Seventy of the past 100 branches already have major banks in their market.

> *It's moving right past banking now – 'what are the big issues within our own community that we are looking to solve?' And they're using these enterprises to be able to coordinate, aggregate the enthusiasm and demand within their own communities to actually solve the problems in the community.*

One other manager suggested the CEM is about microeconomic reform "in their communities or spheres of influence rather than just doing banking". The CEM benefits both the local communities and B&AB. It provides a structure (meeting regulatory requirements) that enables access to capital for small to medium enterprises, and a mechanism to enable 'locals to invest locally'. For B&AB, it allows it to access new markets and expand its business with little capital outlay. A long-term manager explained it as follows:

> *Community bank has been without a doubt the best initiative that our business has taken in my time at the bank. There is no way for instance that we could have funded a distribution network of 450 branches; there is no way that we would have 1.4 million customers now. We wouldn't have the scale, the distribution. It gave us our opportunity to establish our place in the banking sector.*

## Collaboration and cooperation

Jones (1995) argues that a firm will gain competitive advantage from developing relationships with its stakeholders based on mutual trust and cooperation, because it substantially increases the "eligibility to take part in certain types of economic relationships and transactions that will be unavailable to opportunistic firms" (Jones 1995, p. 422). B&AB managers reinforce that mutual trust, cooperation and collaboration are key elements in its community engagement strategy, while opportunistic behaviour can be damaging. They stress that the ability to collaborate is a core strength, and this allows B&AB to build a connection with the customer that is not just price based. B&AB managers believe that a strategy of tying the bank's success to the success of its communities through a collaborative CEM will lead to long-term growth:

> *You will have heard by now 'successful customers, successful communities creates a successful bank' in that order. It's a really important distinction. It isn't chasing a shareholder objective at the expense of customers or communities. It's actually understanding that if we help and improve the financial prospects of the communities within which we work by default it will enhance our business at the backend because without them we don't have a business.*

Without the buy-in from the local communities, the CEM would not succeed. The model requires the communities to "put skin in, in return for a fair share of the reward", and there has to be absolute commitment from the communities and B&AB, "because if you don't have that, then it's not worth trying to pursue [the CEM]". While the structure of the CEM could be easily replicated by competitors, the intent of the relationship is more difficult to replicate:

> *It is a model which is based on commitment, trust, collaboration, respect, accountability as well as sitting between two slices of 1) commercial viability*

*and 2) the strength of relationship. This is a very hard thing for corporates to do because corporates traditionally want to own and take control.*

## Values, leadership and culture

Interviews with B&AB staff revealed that values, leadership and culture are important factors in the success of B&AB's community engagement strategy. Trust and relevance were the values most cited by staff, followed by loyalty, integrity and honesty. Equally if not more important to the success of the CEM is leadership. In the first ten years of the CEM, the managing director Rob Hunt was the key driver of the vision and strategy. In 2009, Rob Hunt retired and Mike Hirst was appointed Managing Director. One staff member explained that with the change in leadership, there has been a reinforcement of the community engagement strategy and a "more heightened commitment to ensuring it continues, accelerates and enhances". In addition, the B&AB Board fully supports the strategy and it has now appointed the chairman of the first community bank to the Board. The values supporting the community engagement strategy are well engrained in the culture, or ethos, of the organisation:

> *CEM is not only a really important part of our business model but its representative of the soul of the place too. Robert Johanson, the Chairman, was speaking at the National Community Bank Conference [in 2010] and he said something like this: 'Even though community bank delivers only about 15% of group profit, it's the defining identity of B&AB'.*

Getting internal buy-in for the CEM was the most difficult challenge early on. Implementing a business model that was very different to the rest of the industry was difficult for some employees to accept in the early days:

> *It's an enormous journey this organisation has gone through in getting used to it. You have your champion teams that press out different alternative ways of thinking of distribution. But at the end of the day it doesn't mean that the organisation comes along. It's many, many years later ... that gradually there is an adoption of it.*

B&AB has put business processes in place to maintain the values and culture that support a community engagement strategy, such as measuring staff on how they model the corporate values. B&AB implemented a balanced scorecard approach where staff performance is measured on five dimensions, each with an equal weighting: customer, community, people, financial and business process. Every B&AB branch is involved in community engagement activities. In addition, B&AB has aligned its organisation structure to embed the CEM in the organisation:

> *... we had an executive that was responsible for the company-owned branches. And then we had the executive that was responsible for community banking. We actually now have one executive that is responsible for retail and it incorporates community and company owned branches all within the one structure now. [The reason for] doing that was twofold. Recognising that it IS a major part of our organisation but it also gave a broader number of staff within the organisation an ability to get involved in the day to day community life.*

# Challenges and opportunities

The global financial crisis (GFC) of 2007–2009 put pressure on access to wholesale funding markets and margins across the industry, and B&AB saw a reduction in its profits. However, B&AB believe that the GFC was a test of the resilience of the CEM, as one executive explained:

> The GFC was the toughest time we have gone through in a long time and we saw a reduction, well and truly, over that period. But it stress-tested that model [and] the organisation came through. I think a number of people, especially the analysts, have turned around and said this is sort of serious now. It does work; it does stand up to that rigorous testing.

Another manager pointed out that the ability to scale is also a challenge – "How do you go from an experiment to a proof-of-concept, to a pilot, to a production and how do you also maintain local commitment" – as well as ensuring the customer experience is always enhanced? In addition, the number of opportunities and requests from communities are increasing, and B&AB has "more needs and opportunities than we have the resources to mobilise for, at this particular point in time".

B&AB believe that the CEM opens up huge opportunities to help communities address their issues in the future. Aged care is one area that B&AB has identified as an emerging issue in communities because of the ageing population in Australia: 'What does the model look like that helps these communities to ensure that they've actually got the facilities and so forth that are required for an ageing population in those areas?' Another opportunity lies with utilising social media. In 2010 B&AB launched an online community networking website <www.planbig.com.au>. The site allows people to start a project and invite others to get involved, or support an existing plan to help make it happen. For example, a plan to set up a local children's playground. One staff member explained:

> This whole piece was around the recognition that social media is such a big part of people's lives now and how do we actually play a relevant role. This is an online example of community engagement, where people actually come together where they've got ideas or plans of things that they'd like to achieve. They help each other out to actually get the outcomes that they're looking for.

# Conclusion

B&AB has taken a very different approach to the traditional shareholder-oriented banking organisations, who typically have a short-term focus on maximising shareholder value. B&AB's strategy is driven by a long-term view of what will make it relevant to its customers and communities – what will make people 'press the button for Bendigo Bank' in the future. While the CEM does adopt the shareholder-ownership structure of the traditional business model, there are significant differences: the CEM shareholders are local community members, not absentee shareholders; eighty per cent of the community enterprises' profits is reinvested back into community development initiatives (twenty per cent is distributed as dividends), and; the CEM is a demand-side model. The model aggregates demand of the community members for essential services and, with this combined buying-power, enables the communities to negotiate better terms and conditions with the suppliers.

In fact, B&AB may not look like a bank in the future, according to one executive:

> *I don't see us looking like a bank in the future, and this isn't the organisation's view, this is my view... To me we're a distribution business, a facilitation business. So I see that those physical networks [will] look a lot different and that's why I tend to use the word shops instead of branches because I think that these physical points out in the communities will be community shops, community sites, where functionally the communities will come together, where the services flow-through, and where things potentially get done. I think, in my mind, things will look very different in the future.*

This style of business requires distributed leadership across networks and organisational boundaries. As noted by the former managing director, "it requires strong leadership, a commitment to education, and the development of skills necessary to establish, manage, and provide appropriate governance for this new business". As the network of community enterprises grow, B&AB may face challenges in terms of managing the complexity of multiple partners and community enterprises, and the issue of power distribution between B&AB and the community businesses has been raised by community banks.

# References

<www.bendigoadelaide.com.au>, viewed 1 September 2010.

Jones, TM 1995, 'Instrumental stakeholder theory: A synthesis of ethics and economics', *The Academy of Management Review*, 20: 404-437.

Putnam, RD, Leonardi, R & Nanetti, R 1993, *Making democracy work: civic traditions in modern Italy*, Princeton, N.J.: Princeton University Press.

Strauss, AL & Corbin, JM 1980, *Basics of qualitative research: techniques and procedures for developing grounded theory*, Thousand Oaks: Sage Publications (first published 1990).

Stubbs, W & Cocklin, C 2007, 'Cooperative, community-spirited and commercial: Social sustainability at Bendigo Bank', *Corporate Social Responsibility & Environmental Management*, 14: 251-262.

SustainableBusiness.com 2009, 'The 2009 SB20: World's Top Sustainable Stocks', <http://www.sustainablebusiness.com/progressiveinvestor/SB20_2009.htm>, viewed 1 September 2010.

# Case 2

# City of Mandurah: Champions of change

André Taylor

## Introduction

Within the Australian water industry, "the current generation of leaders faces a tremendous opportunity and challenge to shape cities of the future as sustainable and liveable places" (Skinner & Young 2010, p. 2). In Australia's southern cities these challenges and opportunities are most apparent. Rapidly increasing populations, high levels of climatic variability, climate change impacting water resources, periodic water shortages, and vulnerable aquatic ecosystems are some of the factors that are driving a transition from traditional methods of managing water to more sustainable ones (Davis 2008; Kaspura 2006; Skinner & Young 2010; WSAA 2007).

The adoption of 'water sensitive urban design' (WSUD) in land developments (e.g. subdivisions), buildings and other infrastructure (e.g. roads) is an expression of this transition. WSUD is defined as "a new paradigm in the planning and design of urban environments that is sensitive to the issues of water sustainability and environmental protection" (Wong 2006, pp. 1–2). The adoption of this paradigm leads to infrastructure such as rain-gardens that filter contaminants from stormwater, and aquifer storage and recovery systems that recharge aquifers with filtered stormwater for subsequent reuse as a secondary water source (Engineers Australia 2006).

In Australia, local government agencies have a major role to play in promoting and implementing WSUD. This is because they are centrally involved in the assessment of new developments and buildings, including the design of stormwater drainage infrastructure. In addition, they design, construct and maintain public infrastructure that can incorporate WSUD, such as roads, parks and reserves. They also commonly maintain WSUD infrastructure that has been constructed by developers, such as estate-scale constructed wetlands and infiltration systems.

The adoption of WSUD is occurring unevenly across Australian cities. Within each state and major city there are regions of activity and inactivity. Typically, regions of activity are characterised by the local leaders at political, executive and/or project levels (change agents) embracing more sustainable approaches and driving change. Such leaders commonly use local planning instruments to require WSUD through new developments, and use pilot projects to demonstrate the practicality of WSUD and resolve technical issues (Brown & Clarke 2007; Mitchell 2004; Taylor 2008).

The role of local leaders who have emerged to champion WSUD and other forms of sustainable water management has been the focus of recent research activity. It is now widely recognised that these champions play a critical role in the transition to 'water sensitive cities' (Brown et al. 2009). The significance of this role has been recognised by academics (Brown et al. 2006; Cashman 2008; Mitchell 2004; Meijerink & Huitema 2010), industry practitioners (Allan et al. 2009; Edwards et al. 2006; Keath & White 2006) and politicians (Commonwealth of Australia 2002).

This chapter presents a case study involving the City of Mandurah local government agency (council) in Western Australia (WA), approximately 70 km south of Perth. It focuses on the role of a project champion (Howell & Higgins 1990a) in 2007, who operated at a middle management level to trigger and drive a range of initiatives to deliver WSUD. At this time, the City of Mandurah was an outstanding and relatively rare example of a local government agency that was making solid progress in the implementation of WSUD in WA.

This case study has three aims. First, it aims to explore the role of change agents in government agencies who initiate and drive processes of influence to deliver more sustainable ways of managing water in cities, with a focus on the project champion role. Second, it seeks to identify the many factors that contribute to the process of change, such as key behaviours and strategies used by the project champion, the role played by other leaders and enabling contextual factors. Third, it highlights some of the practical implications of the case study as a resource to sustainability practitioners who work in similar contexts and wish to be more effective at driving change.

The chapter begins by briefly introducing emergent leaders who have been variously described as change agents, champions and policy entrepreneurs. It then provides a description of the case study organisation and surrounding environment. A three-phase conceptual model is then introduced to help examine the role of change agents in promoting WSUD policies and projects in water agencies. This model is then used to highlight important factors from the case study that existed at the individual, team and contextual level. The chapter concludes by discussing the relevance of the case study to other contexts, and describing three practical implications.

# Background

## Change agents, champions and policy entrepreneurs

A variety of terms in different bodies of literature have been used to describe individuals and groups who initiate and drive change. The term 'change agent' has been broadly defined as "any individual or group that initiates and/or facilitates change" (Duncan 1978, p. 362). Ottaway (1983) developed a taxonomy of ten types of change agents. One of these was the 'key change agent', who is an *initiator* of change processes. Key change agents have advanced leadership skills, choose the right time to initiate change, and their leadership behaviours and values tend to dominate change processes (Ottaway 1983).

Another closely-related term is the 'champion'. Within Australian water agencies, sustainable urban water management (SUWM) champions have been defined as emergent leaders who have specific attributes and are adept at influencing others to adopt SUWM principles and practices, such as WSUD (Taylor 2008 & 2010a). These attributes include the personality characteristics of creativity, persistence and resilience, and values such as a strong personal commitment to SUWM and environmental sustainability. Other examples include a good general knowledge of the water industry and associated technology, advanced skills at exercising influence and high levels of personal power. In addition, they commonly use key leadership behaviours, such as identifying opportunities for influence, choosing the right influence tactics for the right target and time, developing and encouraging colleagues and undertaking advanced forms of social networking (e.g. strategic networking; see Ibarra & Hunter 2007). As acknowledged 'leaders', these individuals also engage in group-based processes of influence that involve establishing direction, aligning resources, and motivating and inspiring their colleagues to implement SUWM.

The 'champions of innovation' literature has highlighted the existence of two types of champion that often exist within an organisation to advance new ideas or products. These are 'project/product champions' and 'executive champions' (Howell & Higgins 1990a; Howell *et al*. 2005; Maidique 1980). Descriptions of project champions are very similar to Ottaway's (1983) description of key change agents. In addition, project champions promote change on a daily basis within organisations or broader institutions, and primarily rely on personal forms of power. In contrast, executive champions are more senior leaders with high levels of position power who allocate resources to innovations and who share some of the associated risks (Maidique 1980). Executive champions often work in tandem with project champions to progress initiatives (Witte 1977).

The term 'policy entrepreneur' is also used in the public policy literature. Policy entrepreneurs are "advocates for proposals or for the prominence of ideas" (Kingdon 1984, p. 129), who "influence policy outcomes by coupling the problem, the politics, and the policy streams when the policy window opens up" (Rains & Prakash 2005, p. 5). In water-related contexts, the term is commonly applied when people, coalitions or organisations use their leadership skills to influence policy and legislation (Brouwer *et al*. 2009; Meijerink & Huitema 2010). Five strategies appear to be critical to their ability to influence policy (Huitema & Meijerink 2009 & 2010). First, they develop or find new ideas. Second, they build coalitions to sell these ideas. Third, they recognise and exploit windows of opportunity (Kingdon 1995) to effect change. Fourth, they find and use multiple venues to promote their ideas. Fifth, they orchestrate networks between stakeholder groups to build shared visions and align resources.

# An overview of the case study organisation and surrounding environment

This section relates to the City of Mandurah local government agency and its surrounding environment as it was in September 2007. It broadly describes the features that may have affected the emergence and effectiveness of a WSUD change agent at that time. An analysis of which contextual factors were particularly beneficial is provided later in the chapter.

In 2007, Mandurah was a regional centre 174km$^2$ in size, 72 kilometres south of Perth with a population of approximately 61,500 (Claydon 2010). The population of this coastal city was growing at a rate of approximately 3.9% per year and the average five year growth rate was 4.4% per year (Claydon 2010), making it one of the fastest growing areas of Australia (Heal 2007). A significant feature of the city was its proximity to waterways that had many important environmental and social values. One such waterway was the Peel Inlet.

Drivers for implementing sustainable forms of water management included rapid population growth, waterways that were under significant stress due to nutrients draining from their catchments (e.g. the Peel Inlet and Harvey Estuary), and an increasing need to conserve surface and groundwater resources. In addition, the region's prosperity and rapid pace of development had created a window of opportunity to ensure WSUD was included in new infrastructure.

The city's natural assets, such as its waterways, were highly valued by the community and elected members (City of Mandurah 2005b). The local Mayor was also a visible and influential advocate (i.e. political champion) for sustainable development. And there was a strong commitment to maintain the city's high levels of liveability. For example, Mandurah was a finalist in the 2006 International Liveable Communities Award, achieving the highest score for the category of 'environmentally sensitive practice'.

The Mandurah region had a long history of water quality problems including severe algal blooms and fish kills in the Peel Inlet-Harvey Estuary system since European settlement, and eutrophication of its rivers and constructed lakes (Heal 2007). Historically, water quality degradation had been so severe in the Peel Inlet-Harvey Estuary system that the State government built the AU$37 million (1994 Australian dollars) Dawesville Channel in the early 1990s to provide greater flushing of the estuary. This history, along with the many environmental values associated with local waterways, helped to create a relatively high level of public awareness concerning water quality, waterway health and the need for improved catchment management.

The City of Mandurah Council was an organisation with 540 staff in 2007. The lead agent in Council for the promotion of WSUD was the Works and Services Directorate. Council's primary responsibility for urban water management related to drainage. In 2007, the WA Water Corporation did not have any main drains in the region, so all drainage issues were managed by Council. The situation was unusual within the greater Perth region. Council had the ability to deliver WSUD assets by regulating proposed developments (e.g. subdivisions and commercial developments), as well as through its own construction activities.

Council also worked cooperatively with the Peel Harvey Catchment Council (Inc.) on a range of water quality and water conservation initiatives. This natural resource management group acted as a 'bridging organisation' (Brown & Clarke 2007) to help build the capacity needed to deliver WSUD throughout the region. For example, in collaboration with the City of Mandurah, it developed model town planning policies and supporting technical guidelines for WSUD that could be used by all local government agencies in the region.

Council's mission statement was "to create a vibrant and sustainable community, maximising opportunities through innovation and partnerships" (City of Mandurah 2005b, p. 3). Its strategic objectives included the promotion of "sustainable urban design", and the protection and enhancement of waterways and coastal areas in the Peel-Harvey region (City of Mandurah 2005a & 2005b). At the project level, in 2007 Council was implementing an International Council for Local Environmental Initiatives (ICLEI) water campaign (City of Mandurah 2007) to improve water quality and conserve water in the region. It had also facilitated: a draft local planning policy for WSUD (2007); an internal policy for implementing WSUD via Council works (2007); and WSUD technical guidelines (2006). Council had also implemented a number of pilot projects for WSUD, such as the award-winning 'Sustainable Mandurah Home' project (City of Mandurah 2005a).

# Methodology

This case study was one of six undertaken as part of a multiple case study research design (Yin 2003). For details of this larger research project, including a more detailed description of the methodology, see Taylor (2010a).

The primary research question was: "What are the main factors that assist project champions to successfully promote sustainable urban water management (SUWM) in Australian publicly-managed water agencies?" The City of Mandurah was selected as a case study agency following consultation with WSUD practitioners (e.g. leading consultants and academics) to find a publicly-managed water agency in WA that was delivering WSUD outcomes and employed a staff member who was initiating and strongly driving WSUD projects. The City of Mandurah was the organisation that was most strongly nominated during this process.

The collection and analysis of data involved seven steps. First, a 2.5 hour semi-structured group interview was held. This interview involved five staff who played different roles in the promotion of WSUD (e.g. development assessment, environmental management and asset management).

Second, as part of the group interview, a questionnaire was used to conduct an anonymous peer nomination process to identify people who played six specific roles in promoting WSUD, including the roles of project and executive champions (Taylor 2010a for a description of these roles). All of the surveyed group interviewees nominated the same person as performing the project champion role.

Third, 1.5 hour semi-structured individual interviews were undertaken with the people most strongly nominated for the six roles. Only four of these interviews were undertaken, as some of these people were nominated for more than one role.

Fourth, a two-part multi-rater questionnaire was administered to the four individual interviewees, as well as their supervisors and five of their peers. Thus, 28 of these questionnaires were administered in total. All were completed and returned. The first part of the questionnaire gathered information on a wide variety of leadership attributes such as personality characteristics, personal values, types of knowledge, leadership behaviours, influence tactics, types of power, social network characteristics, and leadership effectiveness. The second part of the questionnaire was a proprietary instrument, namely, the Multifactor Leadership Questionnaire (MLQ - Form 5X) (Avolio & Bass 2004). This instrument measures the frequency with which leaders, such as champions, use particular types of transformational leadership[2] behaviours (e.g. questioning the status quo and expressing enthusiasm and confidence). This was included as there is strong evidence to suggest that change agents, champions and environmental leaders frequently use transformational leadership behaviours (Ashkanasy & Tse 2000; Egri & Herman 2000; Howell & Higgins 1990a; Howell et al. 2005; Portugal & Yukl 1994).

Fifth, to better understand the context of WSUD-related leadership in the case study agency, an analysis of relevant documents was undertaken (e.g. corporate and strategic plans, water-related action plans and regional waterway management plans). This was supplemented by a two hour interview to gather additional contextual information with a person who was familiar with the organisational unit in which the nominated project champion worked (i.e. the Infrastructure Services Team).

Sixth, a variety of data analysis methods were used, including the use of summary notes and 'memoing' (Miles & Huberman 1994) after each interview, as well as document analysis. In addition, content analysis of the group and individual interviews was undertaken using methodology described by Bryman et al. (1996), Dutton et al. (2001), Howell and Boies (2004), Howell and Higgins (1990a & 1990b), Miles and Huberman (1994), and Patton (1990). Descriptive, role ordered matrices (Miles & Huberman 1994) and summary statistics from the content analysis of interview notes were also used to identify attributes of the project champion that were strong, weak and/or unusual when compared to a local control group (i.e. the people most strongly nominated for the four non-champion roles during the group interview). The small sample sizes prevented meaningful statistical analysis beyond the use of summary statistics for quantitative data. When examining such data from the questionnaire, differences between the project champion's ratings and those from the control group were regarded as being substantial if they were at least half a rating point on the relevant Likert scales.

Seventh, multiple sources and methods were used to identify common themes in the data and to draw conclusions. For example, to make conclusions relating to the

---

[2] "Leadership behaviors that transform and inspire followers to perform beyond expectations while transcending self-interest for the good of the organization" (Avolio et al. 2009, p. 423).

transformational leadership behaviour of the project champion, data were used from the individual interview with the project champion as well as the customised and proprietary portion of the multi-rater questionnaire. This questionnaire generated three groups of ratings (i.e. self, supervisor and average peer ratings).

# The case study

## A conceptual model to help understand the role of change agents

A conceptual model is presented in Figure 2.1 that provides a framework for understanding the process that typically takes place in Australian water agencies when project champions drive new policies and projects to advance more sustainable forms of urban water management. It includes three main phases. This model was derived from a multiple case study (Taylor 2010a), which included the case study being discussed here. This section provides an overview of the model, which will be used to help understand the role of WSUD champions in the case study agency, with a focus on the project champion role.

During the *Initiation* phase of the process shown in Figure 2.1, SUWM project champions typically emerge as leaders as a consequence of their personality traits and context. They frequently use distinguishing leadership behaviours at this time (e.g. specific transformational leadership behaviours). Throughout the challenging *Endorsement* phase, project champions typically work with their peers (i.e. laterally) and executives and politicians (i.e. vertically) to secure approval and/or funding for new initiatives. Relationships are critical at this time, as are windows of opportunity to exert influence. During the final *Implementation* phase, project champions commonly collaborate with colleagues from across organisational boundaries, often in the form of multi-disciplinary teams. The model also highlights that executive and/or political champions often play an 'enabling leadership' role (Uhl-Bien *et al.* 2007) to create supportive environments for SUWM and less senior leaders, such as project champions, to interact, innovate and collectively drive new approaches.

## The Initiation Phase

During this phase, the nominated project champion (hereby referred to as 'PC') stood out as an emergent leader who was strongly motivated to advance WSUD initiatives (e.g. new projects). His personality traits, sources of power, and his surrounding context contributed to his emergence as a leader. In terms of traits, he had relatively strong environmental values which were aligned with the WSUD philosophy. These values were also aligned with sustainability-related values in the organisation's culture, the Mayor's strong personal commitment to sustainability, and the values of a small group of colleagues with whom PC often collaborated to drive WSUD projects. This issue of value alignment is a common feature of environmental leadership (Boiral *et al.* 2009).

**Figure 2.1** A three-phase conceptual model of champion-driven leadership involving the promotion of sustainable urban water management (SUWM) initiatives in Australian water agencies

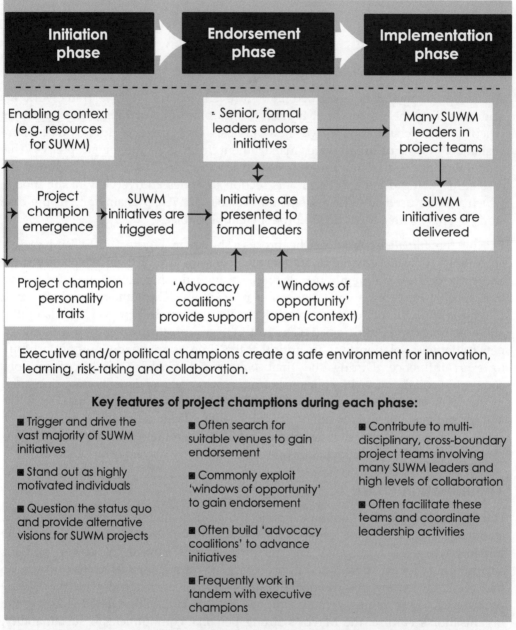

*Source*: Modified from Taylor, A, Cocklin, C, Brown, R & Wilson-Evered, E (in press), 'An investigation of champion-driven leadership processes', *The Leadership Quarterly*.

Other personality traits that assisted PC to initiate new ideas and projects during this phase included relatively high levels of openness to experience, propensity to take risks, persistence, motivation, enthusiasm, energy, self-sufficiency and optimism. He also had relatively low levels of agreeableness[3], helping PC to question the status quo. Relatively high levels of the 'openness to experience' trait are also common in WSUD project champions. In relation to this trait, PC stated that: "… I'm always forging change in practices and the way we do things … always trying to push it, do it differently and take a few risks".

PC had relatively high levels of personal and position power. His WSUD-related technical expertise and strong relationships with colleagues were two important sources of personal power. His relatively senior position in council (i.e. two tiers below the Chief Executive Officer) and his role in the development assessment process generated considerable position power. This power created opportunities to exert influence. For example, in relation to the importance of PC's power as a regulator of new, private-sector development, he stated: "It opens doors and gives you opportunities. So, I suppose we just try and use those power bases as a way of creating change".

The context surrounding PC included many enabling factors which helped PC to emerge and be effective as a WSUD champion. These factors included strong support from the nominated executive champion (hereby referred to as 'EC'), and the activities of the Mayor who championed sustainable development in the region. In addition, the organisation had an enabling culture that encouraged innovation, learning, adaptive management, responsible risk-taking, collaboration, sustainability, and the empowering of staff to make decisions. A small, cross-boundary group of committed staff who shared a common commitment to sustainable development was another important enabling factor. It also helped that the local economy was strong, there was understanding across the community of the need to protect the health of local waterways, and the organisation had a robust policy framework for sustainable development. The relatively small size of the organisation also helped PC to build strong, mutually beneficial relationships with key staff across the organisation's functional boundaries to help drive change.

The relationship between PC and EC was significant for several reasons. First, EC helped to mentor PC and guide his development as a WSUD leader. Second, EC provided support to PC when there was some resistance to change. For example, PC described EC as: "… a good supporter, and I know that if I've got the right idea and our office people aren't seeing it and there's some negativity, I know he will support the change anyway …"

Third, EC sometimes played a mediation role between PC, who strongly drove change, and people in the organisation who were more resistant to change. Such behaviours are consistent with EC being an enabling leader (Uhl-Bien *et al.* 2007). Such leaders often have the challenging role of working at the interface between lower-level emerging leaders, such as PC, who drive transformative change from

---

[3] Agreeableness levels were relatively low compared to control groups, but still moderate (e.g. for questionnaire data, around 4.5 on a 1-7 Likert scale). Consequently, PC was cooperative in nature but was also comfortable about proposing alternatives to traditional ways of managing water that were not supported by all stakeholders.

the bottom-up, and more senior leaders who typically are more comfortable with incremental, top-down change.

PC was a relatively strong transformational leader (Bass 1985) who frequently used 'inspirational motivation' behaviours (e.g. expressing enthusiasm and confidence). This is common amongst SUWM project champions (Taylor 2010a) and other types of champions (Howell & Higgins 1990a).

Other frequently used behaviours during the *Initiation* phase included articulating visions for projects, communicating clearly and frequently, and questioning the status quo. In addition, PC commonly tailored communication activities for specific audiences, planted the seeds of ideas with stakeholder groups, and prepared for anticipated criticisms of new ideas. In particular, persisting under adversity and providing inspiration and motivation to others were highly relevant to PC.

## The Endorsement Phase

During the *Endorsement* phase, there was evidence that PC undertook 'venue shopping' behaviours (Huitema & Meijerink 2009) to identify opportunities to market WSUD, promote WSUD initiatives and drive change. The primary venue PC used to exert influence was within the formal development assessment process because he had significant position power in this arena. He also sought to encourage the adoption of WSUD within the organisation (e.g. via council's own works) and in the region (e.g. through the Peel-Harvey Catchment Council). Given PC's ability to strongly drive change through the development assessment process, many of the 'senior formal leaders' (see the conceptual model shown in Figure 2.1) that PC influenced were property developers in the private sector – as opposed to executives in council and councillors. In this context, PC could often use his position and personal power to exercise influence without the need for substantial assistance from other people or groups. However, for initiatives outside the development assessment process (e.g. establishing new pilot projects), gathering support from colleagues and building coalitions became more important.

PC was a good networker who showed a preference for the 'strong tie' strategy of social networking[4]. Given his work context, PC relied more heavily on relationships with colleagues (operational networks; see Ibarra & Hunter 2007) than relationships with council executives and politicians (strategic networks) to drive change. This may help to explain why PC was more successful at promoting the adoption of WSUD via new development in the private sector than through Council's own works. SUWM project champions in other water agencies who focus their influence attempts *within* their organisations typically have unusually strong strategic networks with executives (including executive champions) and local politicians to help them 'influence upwards' during the *Endorsement* phase (Taylor 2008 & 2010a).

There were a variety of windows of opportunity that helped PC to foster WSUD. The rapid pace of development in the Mandurah region, combined with PC's formal

---

[4] This strategy involves developing a small number of mutually beneficial relationships with colleagues who have a high degree of centrality in other social networks (see Granovetter 1973).

role in the development assessment process created many development proposals that could include WSUD. Another window of opportunity was created by Council executives actively managing the organisational culture using tools such as the 'Organisational Culture Inventory' and 'Leadership Styles Inventory' (Jones *et al.* 2006). This had been occurring for two years and had produced a culture that was relatively conducive to the philosophy of WSUD (e.g. it valued innovation, learning, adaptive management, collaboration and sustainability). The advocacy of the Mayor also influenced the organisation's culture and policies in a way that helped to support WSUD-related initiatives. Both the Mayor and those executives involved in the management of the organisation's culture were engaged in forms of 'enabling leadership' that indirectly assisted PC and his colleagues to advance WSUD initiatives.

Key behaviours used by PC during this phase included using pilot projects to demonstrate the feasibility of WSUD (e.g. to sceptical developers), taking opportunities to market WSUD in a range of forums, and building coalitions of support within Council and the region. PC showed a preference for working on a one-to-one basis or in small groups of people. For example, PC invested the time to develop constructive working relationships with local developers, sometimes taking them on personalised tours of recently built WSUD assets. The strength of PC's relationships with others (e.g. a small group of colleagues in the organisation who shared similar values) was one key to his ability to successfully deliver WSUD-related projects.

## The Implementation Phase

During the *Implementation* phase, PC commonly worked with colleagues across organisational boundaries who shared a common 'care factor' for sustainable development to deliver projects and policies. Interviews with staff indicated a high level of trust existed between these colleagues, which helped collaboration. This small, informal team of committed professionals is an example of coordinated 'distributed leadership' (Gibb 1954)[5], with PC sometimes coordinating the activities of other leaders in the team.

Behaviours PC used that were beneficial during this phase included working with stakeholders to facilitate outcomes, getting the right people involved at the right time with projects, and collaborating with others. His leadership style was usually collaborative and flexible during all phases. Even in the context of assessing development proposals, PC sought to collaborate to deliver good WSUD outcomes. In this context he stated: "We will discuss what we'd like to see and 'vision it' and 'workshop it' even as the proposal is developed. So it's more of a collaborative approach to a project".

There was also evidence of some conflict between PC and others who resisted change, particularly within council. Such conflict appeared to stem from differences in people's personal values and openness to change, as well as PC's strong desire to

---

[5] The distributed leadership model (Gibb 1954) conceptualises leadership as a process of influence that occurs within groups and involves more than one leader.

deliver on-the-ground outcomes. PC was aware of the potential for interpersonal conflict, and was seeking to modify some of his behaviours to better engage others. In this context, he stated:

> I probably get a bit negative with the 'knockers' ... it's probably not the best but I might be a bit driven to make the change and I often come against [people] - even people internally here - where they don't quite see it the same way I do. ... I suppose I'm always trying to get through to the point at the end and maybe I need to be more mindful of others' comments along the way.

A feature of this case study was the ability of PC to operate effectively in all three phases of typical WSUD-related leadership processes (see Figure 2.1). During the *Initiation* phase, he was able to trigger ideas and potential projects as a relatively individualistic leader. During the *Endorsement* phase, he was able to identify venues where it could exert influence, collaborate with others to get agreement on initiatives and use windows of opportunity. For example, he was able to work collaboratively with developers to incorporate WSUD into the design of new developments. During the *Implementation* phase, he was able to collaborate with others to deliver projects, despite evidence of some interpersonal conflict with those who resisted change. In particular, he often worked with a small, informal team of committed professionals who worked across Council's functional units to deliver major projects.

## Relevance of the case study to other contexts

Although more research is needed to examine the attributes of effective environmental champions in a broader range of contexts, available research indicates that such leaders share many similarities. For example, many of the attributes of environmental leaders described by the corporate sustainability literature on change agents, the policy science literature on policy entrepreneurs and the environmental management literature on champions are the same (see Dunphy *et al.* 2007; Meijerink & Huitema 2010; and Andersson & Bateman 2000 respectively). These bodies of literature emphasise personality characteristics such as confidence, persistence, enthusiasm and motivation. Personal values such as a strong commitment to continuous learning are also commonly reported. They also highlight the importance of the leader having an excellent general knowledge of their organisation, industry and institutional system. In terms of key behaviours, they indicate these leaders adeptly use windows of opportunity and build coalitions of support. They also excel at social networking and creating, manipulating or 'shopping' for venues in which they can effect change. They also frequently use transformational leadership behaviours, such as questioning the status quo, and usually thrive in contexts characterised by significant change.

Available research also indicates some differences between emergent environmental leaders which reflect their leadership context and role. For example, political knowledge appears to be more important for champions working in public sector organisations, such as policy entrepreneurs (Meijerink & Huitema 2010). Also, the tandem model of championship (Witte 1977) is a commonly reported phenomenon. This model highlights how executive and project level champions in organisations can use different leadership behaviours as they work in concert to progress innovations.

In sum, available evidence suggests that many of the key attributes of environmental champions who work in different contexts appear to be similar. Nevertheless, as each institutional setting is unique and leadership is sensitive to context, one would expect some differences between these champions. Thus, it is hypothesised that most, but not all of the strategies used by the project level champion described in this case study would be relevant to other environmental champions working in other contexts. In addition, it is suggested that the case study would be most relevant to environmental champions working in government agencies in western countries.

## Practical implications

This section explores three practical implications from the case study. First, knowledge of the leadership behaviours and strategies that champions use in different contexts can be used to help nascent leaders working in similar contexts to build important skills and learn when to use particular behaviours. For example, the three phase model of typical champion-driven leadership processes in water agencies (Figure 2.1) helps to understand which project champion behaviours are important at different times. It also highlights the importance of executive and political leaders who engage in enabling leadership to directly or indirectly help champions at the project level. This type of knowledge can feed into development initiatives, such as tailored leadership development programs (Taylor 2010b), training courses and mentoring programs.

Second, this case study described several key behaviours that were associated with PC's ability to exert influence and drive WSUD. These included his frequent use of transformational leadership behaviours, which is a common feature of champions of innovation and environmental leaders (Howell & Higgins 1990a; Portugal & Yukl 1994). He also had relatively high levels of position and personal power. In particular, his role in the development assessment process created position power and opportunities to exert influence in the private sector. Multiple case study research has found that the *combination* of personal and position power is broadly correlated with champion effectiveness (Taylor 2010a). PC also invested time in building strong relationships with a small group of committed professionals across the organisation with similar values, then collaborating with these colleagues to drive WSUD-related projects. Other studies have also emphasised the importance of cross-boundary teams and networks of change agents in promoting more sustainable practices (Benn *et al.* 2006; Brown & Clarke 2007). PC also had the knowledge, skills, networks and sources of power to operate effectively in the *Initiation*, *Endorsement* and *Implementation* phases of the idealised leadership process shown in Figure 2.1. Broader studies indicate this is relatively rare amongst project champions driving WSUD in the water sector, and is broadly correlated with champion effectiveness (Taylor 2008 & 2010a).

Third, at the time the case study research was undertaken the context in and around the City of Mandurah was strongly supportive of WSUD. Factors such as the strong economy, sensitive waterways, a history of water-related crises, the community's close association with waterways, and the Mayor's passion for sustainable development helped PC and his colleagues to advance WSUD at a policy and project level. In particular, the concept of 'enabling leadership' is a feature of this case study. The local Mayor provided strong political leadership. Her support for sustainability had a ripple effect that moved through Council's policies, organisational culture, plans and practices. At the executive level within Council, leaders actively managed the organisation's culture to support innovation, learning, adaptive management, collaboration and sustainability. EC provided strong support to PC in the form of mentoring, backing good ideas, and managing the tension that occasionally occurred between PC and colleagues who resisted change. Research in other water agencies has found that strongly driven, independent, risk-taking project champions are more likely to make significant mistakes where they do not have guidance from experienced mentors and/or executive champions (Taylor 2010a). Thus, the nature of the relationship between project and executive champions is important, even for project champions with relatively high levels of position power, as in this case study.

## Conclusions

This case study investigated the role of a champion (a type of change agent) who operated at the project level to drive more sustainable forms of urban water management in the City of Mandurah, Western Australia, in 2007. This local government agency had a strong reputation in Western Australia as being a leader in championing water sensitive urban design (WSUD). The research involved group and individual interviews, an anonymous peer nomination process to identify people playing particular leadership roles, a multi-rater questionnaire and document analysis.

Although the research focused on an individual leader (i.e. a project champion – PC), it found that, in terms of delivering outcomes, the surrounding leadership context and the activity of other leaders were just as important as PC's individual abilities. For example, PC benefited indirectly and directly from the behaviours of: the local Mayor, who was a strong advocate for sustainability; executives within the organisation, who actively managed the organisational culture; an enabling executive champion (EC), who supported and mentored PC; and a small team of committed professionals across the organisation's functional boundaries who shared similar values.

Analysis of influential factors that contributed to PC's ability to drive change was structured around a three-phase conceptual model (see Figure 2.1). This model highlighted the phases that typical WSUD-related initiatives (e.g. policies and projects) typically move through when triggered by a project champion in Australian water agencies. The model helped to understand why PC used certain leadership behaviours, types of power and networks at different times. For example, during the *Initiation* phase, PC was a relatively individual leader, using transformational leadership behaviours like articulating visions for projects and questioning the status quo. During the *Endorsement* phase, he used windows of opportunity to secure agreement for change, such as opportunities that emerged through Council's development assessment process. Finally, during the *Implementation* phase, he often collaborated with others to deliver WSUD projects, sometimes in the form of an informal, cross-boundary, multi-disciplinary team of colleagues.

The case study highlighted several keys to success. These included the highly supportive context in and around the City of Mandurah (e.g. a development boom and highly-valued, sensitive local waterways) which helped PC and other leaders to advance WSUD initiatives. In addition, enabling leadership by the Mayor and executives (including EC), helped to produce policies and an organisational culture that broadly supported WSUD and PC's activities. In particular, there was alignment between values held by PC, the Mayor and a small group of colleagues who commonly worked with PC, as well as the organisational culture. PC also had relatively high levels of personal and position power. His role in the development assessment process was a valuable source of position power which frequently created opportunities to exert influence, in particular during the *Endorsement* phase (see Figure 2.1). As is the case amongst champions of innovation and environmental leaders, PC was a strong transformational leader in the context of promoting WSUD. As such, he frequently used behaviours such as expressing enthusiasm and confidence which provided others with inspiration and motivation. PC also had the skills, knowledge, power and networks to operate effectively in all three phases of typical WSUD-related leadership processes. This ability is relatively rare, and is broadly correlated with leadership effectiveness amongst WSUD champions in water agencies (Taylor 2010a).

The knowledge generated by studying change agents, like PC, and their leadership context has practical value. It can be used to help aspiring change agents to become more effective by understanding the importance of key skills, using certain behaviours at the right time, planning for and using windows of opportunity, as well as building particular types of social networks, power and knowledge. Such knowledge can be packaged in the form of guidelines, training courses and customised leadership development programs (Taylor 2008). Such interventions can then be used by enabling leaders to help grow the next generation of project champions who have the ability to initiate and successfully drive more sustainable practices.

# References

Allan, A, Barich, N, Brennan, M, Catchlove, R, DeSilva, N, Duncan, L, Ewert, J, Hardy, M., Holmes, L, Kaye, E, Kinnear, L, Lee, A, Tay, J, van de Meene, S & Walker, S 2009, *A Vision For A Water Sensitive City*, viewed 13 August 2010, <http:// www.watersensitivecities09.com>.

Andersson, L & Bateman, T 2000, 'Individual environmental initiative: Championing natural environmental issues in US business organizations', *Academy of Management Journal, 43*(4): 548-570.

Ashkanasy, N & Tse, B 2000, 'Transformational leadership as management of emotion: A conceptual review', in N Ashkanasy, C Härtel, & W Zerbe (eds), *Emotions in the Workplace: Research, Theory, and Practice* (pp. 221–235), Quorum Books, Westport, Connecticut.

Avolio, B & Bass, B 2004, *Multifactor Leadership Questionnaire: Sampler Set: Manual, Forms and Scoring Key,* Mind Garden Inc., Menlo Park, California.

Bass, B 1985, *Leadership and Performance Beyond Expectations,* Free Press, New York, New York.

Benn, S, Dunphy, D & Griffiths, A 2006, 'Enabling change for corporate sustainability: An integrated perspective', *Australasian Journal of Environmental Management,* 13(2006): 156-165.

Boiral. O, Cayer, M & Baron, C 2009, 'The action logics of environmental leadership: A developmental perspective', *Journal of Business Ethics,* 85(2009): 479-499.

Brouwer, S, Huitema, D & Biermann, F 2009, 'Towards adaptive management: The strategies of policy entrepreneurs to direct policy change' in *Proceedings of the Conference on the Human Dimensions of Global Environmental Change,* 2-4 December 2009, Amsterdam, The Netherlands, viewed 13 August 2010, <http://www.earthsystemgovernance.org>.

Brown, R & Clarke, J 2007, *Transition to Water Sensitive Urban Design: The Story of Melbourne, Australia,* Facility for Advancing Water Biofiltration and National Urban Water Governance Program, Monash University, Melbourne, Victoria.

Brown, R, Keath, N & Wong, T 2009, 'Urban water management in cities: Historical, current and future regimes', *Water Science & Technology,* 59(5): 847-855.

Brown, R, Mouritz, M & Taylor, A 2006, 'Institutional capacity' in T Wong (ed), *Australian Runoff Quality: A Guide to Water Sensitive Urban Design* (pp. 5-1 – 5-20), Engineers Australia, Melbourne, Victoria.

Bryman, A, Stephens, M & á Campo, C 1996, 'The importance of context: Qualitative research and the study of leadership' *The Leadership Quarterly,* 7(3): 353-370.

Cashman, A 2008, 'Institutional responses to urban flooding: case studies from Bradford and Glasgow' in *Proceedings of the 11th International Conference on Urban Drainage,* 1-5 September, Edinburgh, Scotland.

City of Mandurah, 2005a, *Annual Report 2005-2006,* City of Mandurah, Mandurah, Western Australia.

City of Mandurah, 2005b, *Community Charter and Strategic Plan 2005-2008,* City of Mandurah, Mandurah, Western Australia.

City of Mandurah, 2007, *ICLEI Water Campaign: Corporate and Community Local Action Plan,* City of Mandurah, Mandurah, Western Australia.

Claydon, A 2010, personal communication 29 September 2010.

Commonwealth of Australia, 2002, *Inquiry into Australian Management of Urban Water,* Commonwealth of Australia, Canberra, Australian Capital Territory.

Davis, C 2008, 'Sustainable urban water systems', *Water,* 35(7): 44-48.

Duncan, W 1978, *Organizational Behaviour,* Horton Mifflin, Boston, Massachusetts.

Dunphy, D, Griffiths, A & Benn, S 2007, *Organizational Change for Corporate Sustainability,* 2nd edn, Routledge, London, England.

Dutton, J, Ashford, S, O'Neill, R & Lawrence, K 2001, 'Moves the matter: Issue selling and organizational change', *Academy of Management Journal*, 44(4): 716-736.

Edwards, P, Holt, P & Francey, M 2006, 'WSUD in local government: Implementation guidelines, institutional change and creating an enabling environment for WSUD adoption, in *Proceedings of the 7th International Conference on Urban Drainage Modelling and the 4th International Conference on Water Sensitive Urban Design*, 2-7 April 2006, Melbourne, Australia.

Egri, C & Herman, S 2000, 'Leadership in the North American environmental sector: Values, leadership styles, and contexts of environmental leaders and their organizations', *Academy of Management Journal*, 43(4): 571-604.

Engineers Australia, 2006, *Australian Runoff Quality: A Guide to Water Sensitive Urban Design*, Engineers Australia, Melbourne, Victoria.

Gibb, C 1954, 'Leadership', in G Lindzey (ed), *Handbook of Social Psychology* (Vol. 2, pp. 877–917), Addison-Wesley, Reading, Massachusetts.

Granovetter, M 1973, 'The strength of weak ties', *American Journal of Sociology*, 78(1973): 1360– 1380.

Heal, G 2007, *Growing the City Sustainably: Incorporating WSUD into the Management of Stormwater*, presentation to the International Council for Local Environmental Initiatives, Perth, Western Australia, viewed 3 September 2007, <http://www.mandurah.wa. gov.au>.

Howell, J & Boies, K 2004, 'Champions of technological innovation: The influence of contextual knowledge, role orientation, idea generation, and idea promotion on champion emergence', *The Leadership Quarterly*, 15(2004): 123-143.

Howell, J & Higgins, C 1990a, 'Champions of technological innovation', *Administrative Science Quarterly*, 35(1990): 317-341.

Howell, J & Higgins, C 1990b, 'Leadership behaviors, influence tactics and career experiences of champions of technical innovation', *The Leadership Quarterly*, 1(4): 249-264.

Howell, J, Shea, C & Higgins, C 2005, 'Champions of product innovations: Defining, developing, and validating a measure of champion behavior', *Journal of Business Venturing*, 20(2005): 641-661.

Huitema, D & Meijerink, S (eds) 2009, *Water Policy Entrepreneurs: A Research Companion to Water Transitions Around The Globe*, Edward Elgar Publishing, Camberley, United Kingdom.

Huitema, D & Meijerink, S 2010, 'Realizing water transitions: the role of policy entrepreneurs in water policy change', *Ecology and Society*, 15(2): 1-26.

Ibarra, H & Hunter, M 2007, 'How leaders build and use networks', *Harvard Business Review*, 85(1): 40-47.

Jones, Q, Dunphy, D, Fishman, R, Larné, M, & Canter, C 2006, *In Great Company: Unlocking the Secrets of Cultural Transformation*, Human Synergistics Australia, Sydney, New South Wales.

Kaspura, A 2006, *Water and Australian Cities: Review of Urban Water Reform*, Institute of Engineers Australia, Canberra, Australian Capital Territory, viewed 17 October 2006, <http://www.engineersaustralia.org.au>.

Keath, N & White, J 2006, 'Building the capacity of local government and industry professionals in sustainable urban water management', in *Proceedings of the 7th International Conference on Urban Drainage Modelling and the 4th International Conference on Water Sensitive Urban Design*, 2-7 April 2006, Melbourne, Australia.

Kingdon, J 1995, *Agendas, Alternatives and Public Policies*, 2nd Edition, Harper Collins, New York, New York.

Maidique, M 1980, 'Entrepreneurs, champions, and technological innovation', *Sloan Management Review*, 21(2): 59-76.

Meijerink, S & Huitema, D 2010. 'Policy entrepreneurs and change strategies: Lessons from sixteen case studies of water transitions around the globe', *Ecology and Society*, 15(2): 1-21.

Miles, M & Huberman, A 1994, *Qualitative Data Analysis*, 2nd edition, Sage, Thousand Oaks, California.

Mitchell, G 2004, *Integrated Urban Water Management. A Review of Current Australian Practice*, Australian Water Association and CSIRO, Melbourne, Victoria.

Ottaway, R 1983, 'The change agent: A taxonomy in relation to the change process', *Human Relations*, 36(4): 361-392.

Patton, M 1990, *Qualitative Evaluation and Research Methods*, 2nd edition, Sage, Newbury Park, California.

Portugal, E & Yukl, G 1994, 'Perspectives on environmental leadership', *The Leadership Quarterly*, 5(3/4): 271-276.

Raines, S & Prakash, A 2005, 'Leadership matters: Policy entrepreneurship in corporate environmental policy making', *Administration and Society*, 37(1), 3-22.

Skinner, R & Young, R 2010, *Outcomes From the 'Cities of the Future' Workshop*, Held at Ozwater'10 in Brisbane 9-10 March 2010, viewed 13 August 2010, <http://www.wsaa.asn.au>.

Taylor, A 2008, *Leadership in Sustainable Urban Water Management: An Investigation of the Champion Phenomenon*, Industry report, National Urban Water Governance Program, Monash University, Melbourne, Victoria, viewed 10 November 2008, <http://www.urbanwatergovernance.com>.

Taylor, A 2010a, *Sustainable Urban Water Management: The Champion Phenomenon*, PhD eThesis. National Urban Water Governance Program, Monash University, Victoria, Melbourne, viewed 6 September 2010, <http://arrow.monash.edu.au/vital/access/manager/ Index>.

Taylor, A 2010b, 'Building leadership capacity to drive sustainable water management: The evaluation of a customised program', *Water Science & Technology*, 61(11): 2797–2807.

Taylor, A, Cocklin, C, Brown, R & Wilson-Evered, E (in press; accepted 15 February 2010),
'An investigation of champion-driven leadership processes', *The Leadership Quarterly*.

The Sustainable Mandurah Home, 2007, *Water Information Sheet*, viewed 3 September 2007, <http://www.sustainablemandurah.com.au>.

Uhl-Bien, M, Marion, R & McKelvey, B 2007, 'Complexity leadership theory: Shifting leadership from the industrial age to the knowledge era', *The Leadership Quarterly,* 18(2007): 289-318.

Water Services Association of Australia (WSAA), 2007, *The WSAA Report Card 2006/07: Performance of The Australian Urban Water Industry and Projections for the Future,* WSAA, Melbourne, Victoria.

Witte, E 1977, 'Power and innovation: A two center theory', *International Studies of Management and Organization,* 7(1): 47-70.

Wong, T 2006, 'Introduction', in T. Wong (ed), *Australian Runoff Quality: A Guide to Water Sensitive Urban Design* (pp. 1-1 – 1-9), Engineers Australia, Melbourne, Victoria.

Yin, R 2003, *Case Study Research: Design and Methods,* 3rd edn, Thousand Oaks, California, Sage Publications.

# Case 3

# Fuji Xerox Australia Eco-manufacturing Centre[6]: A case study in strategic sustainability

SUZANNE BENN, DEXTER DUNPHY AND TAMSIN ANGUS-LEPPAN

## Introduction

Fuji Xerox is a leader in the development and application of sustainable operations. A key aspect of the company's leadership is in its ground-breaking development of remanufacturing. Before 1993, broken or damaged parts from Fuji Xerox equipment were sent to landfill and replaced with imported parts. This meant that, for example, a AU$10,000 circuit board with only minor defects was considered waste. This system carried considerable financial and environmental costs. In Australia, it was decided to roll out a trial remanufacturing project that had begun in the US. The Fuji Xerox Australia Eco-manufacturing Centre (the Centre) was established, with the mandate to develop technological capabilities to enable remanufacturing. The Centre now accounts for 80% of Fuji Xerox Australia's spare parts requirements – these parts would otherwise have gone to landfill. The success of the Centre rests on both technological advances and a new, high performance workplace culture. In this case, we explore the challenges faced in developing strategic sustainability at Fuji Xerox.

## Research methodology

The authors have studied the operation of the plant in detail and at first hand over a period of eight years. In that time they have interviewed both senior executives and supervisors and accessed published material and internal documents.

---

[6] The information in this case study was obtained from in-depth interviews with key respondents from the Eco Manufacturing Centre, including the General Manager and Communications Manager, along with direct observations in the plant, and informed by secondary sources such as the Fuji Xerox Sustainability Report 2010 and previous publications by the authors (Benn & Dunphy 2004; Dunphy *et al.* 2007).

# Background

"Remanufacturing was the beginning of the story – in order to foster support for the wider business plan for total product responsibility" (Dan Godamunne, General Manager Fuji Xerox Eco-Manufacturing Centre)[7].

Fuji Xerox was established in 1962 as a joint venture between Fuji Photo Film Co Ltd of Japan and Rank Xerox Limited of the United Kingdom. Fuji Xerox Australia (FXA) is a wholly-owned subsidiary of Fuji Xerox Asia Pacific. Figure 3.1 summarises key statistics including the core values at FXA.

**Figure 3.1** Fuji Xerox Australia key statistics

**Fuji Xerox Australia**

**Managing Director**
Nick Kugenthiran

**2008–09 Revenue**
$803.40 million

**Employees**
1846

**Headquarters**
Level 1, 101 Waterloo Road,
Macquarie Park NSW 2113

**Founded in** Australia in 1960

**Ownership** Fuji Xerox Australia is a wholly owned subsidiary of Fuji Xerox Asia Pacific. Fuji Xerox of Japan owns Fuji Xerox Asia Pacific and is a joint venture between Fuji Film Holdings (75 %) and Xerox Corporation (25 %).

*Source*: Fuji Xerox Australia 2010.

Fuji Xerox produces a wide range of products designed to manage electronic or paper documents, referring to themselves as 'The Document Company'. The company initially began renting its products to customers but later moved on to selling and leasing products with service contracts to repair or replace consumables. This ongoing relationship with each customer subsequently facilitated remanufacturing, making it possible to retrieve used equipment and parts for the process.

The remanufacturing operation in Sydney began in 1993. It was led by a group of three employees, located at Mascot, who had backgrounds in marketing, engineering and service engineering for Fuji Xerox office equipment. Before 1993, broken or damaged parts from Fuji Xerox equipment would be replaced with new

---

[7] Dexter Dunphy and Suzanne Benn, Interview with Dan Godamunne, General Manager Fuji Xerox Eco-Manufacturing Centre, 11 February, 2009.

ones at the Mascot operation. Replacement parts were imported from the US, Japan or Europe.

The project devised by this small group of entrepreneurs was to turn the whole process around, so that worn and failed items, rather than being treated as waste, were redefined as potentially valuable resources for the new process of eco manufacturing. The project has been so successful that failed and exhausted items such as fuser rollers, originally imported into Australia from the Fuji Xerox Asia Pacific region and Japan, are now remanufactured in Australia and re-exported in 'as new' or better condition. The original remanufacturing team of three has grown to a staff of 100 and all, except administrative staff, are directly involved in the eco manufacturing process. Fuji Xerox has invested in the new plant now located at Zetland to the south of the Sydney Central Business District. The Eco Manufacturing Centre has developed into a relatively small but strategically important part of the Australian operations. According to the 2010 FXA Sustainability Report, over the ten years of its operations, the Eco Manufacturing Centre has achieved a AU$240 million return on a AU$22 million investment.

## Eco manufacturing as a business strategy

Fuji Xerox's environmental vision is "*Sustaining our environment through better technology*". Worldwide, the company is committed to developing remanufacturing processes and, for newly designed models, ensuring all consumable parts can be remanufactured. The company has won several US environmental awards for their innovative approaches to 'manufacturing for remanufacture'. In Australia, Fuji Xerox has received numerous awards; most recently FXA was selected as a finalist for the Leading in Sustainability– Setting the Standard for Large Organisations award at the 2010 Banksia Awards (Fuji Xerox Sustainability Report 2010). The aim is to develop fully sustainable products and business processes.

Eco manufacturing takes used components, tests them and then re-engineers and reassembles them into 'new' products while ensuring that the production process and the products do not have adverse environmental effects. However, to produce a re-engineered product as good as, or better than new, while meeting the new sustainability challenges, requires addressing some complex technological challenges. For example, the materials of the components may have changed during their first use. These changes occur due to heat, vibration or some other physical effect of the operational processes within the equipment. So remanufacturing cannot simply replicate the original production process. In addition, in order to qualify as eco-manufacturing the remanufacturing process may have to meet new and higher environmental safety standards.

Fuji Xerox managers describe the work of their Eco Manufacturing Centre as re-engineering and redesigning a product or product component and developing it to equal or better than new condition. Figure 3.2 outlines the basic process involved in remanufacturing. This process involves examining technical causes for failure while looking for opportunities to extend the life of the product and in general improve its performance. These processes also have environmental benefits by reducing the demand for raw materials, energy and waste to landfill. Another major benefit to the business is the acquisition of data about problems that develop in its products over time. Sending used products to landfill meant that these data were lost. Part of the remanufacturing/re-engineering process involves analysing the defects in the components that have been returned. This analysis provides information that can be used to improve component design and thereby leads to the production of 'as new' or better remanufactured products. Hence the business benefits from this approach to manufacturing include

- decreased costs due to recycling

- improved design for increased reliability and enhanced performance

- savings from import substitution and new export earnings.

**Figure 3.2** The remanufacturing and design cycle

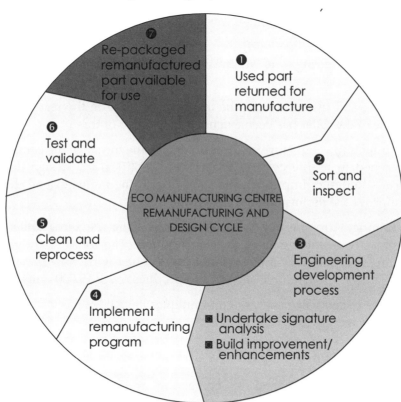

*Source*: Fuji Xerox 2000.

Part of the challenge of creating an effective remanufacturing operation is that it demands expertise in a range of specialised engineering disciplines including optical systems, modulators and sensors, electrostatic and electromagnetic systems, electronics and material sciences. But recurrent faults often occur because of the interactions across these disciplinary boundaries. As a result, finding innovative solutions may require an integrated diagnosis and a holistic redesign solution. The challenge is to maintain depth of expertise in specialised areas while developing high levels of collaboration in creating innovative solutions to identified problems.

# The eco manufacturing technical system

## Eco-manufacturing operations

Fuji Xerox's Eco Manufacturing Centre conducts six production operations involving fuser rollers, laser optical systems, electronics, magnetic rollers, mechanical assemblies and xerographic module remanufacture. The operations involve:

- stripping the fuser roller, a vital component of any multifunctional device, and recoating it to exacting specifications that result in high quality prints and longer operational life. The patented process developed for this is so successful that remanufactured rollers are superior to the originals

- testing, repairing and realigning the laser optical system. This was not feasible until the technology was developed at the Centre. Remanufactured laser assemblies are produced for the local market and for export

- testing, repairing and upgrading the electronic circuitry to improve reliability

- testing and re-engineering the magnetic roller with the result that the life cycle of the roller has been extended by a factor of three

- conducting a 'signature analysis' to determine the stage the component has reached in its life cycle. This process identifies any design defects and leads to design improvements to overcome the faults, extend the life and improve the performance of the product

- cleaning, stripping down, checking and testing the xerographic modules before reassembling them as new products. The toner cartridges are not simply refilled, they are remanufactured. The Centre now remanufactures 175 major sub-assemblies/parts and produces over 15,000 units per month.

The innovative remanufacturing design techniques have been passed back into new product designs worldwide.

## Component recovery

This new remanufacturing and re-engineering approach involved Fuji Xerox managers rethinking a significant part of their business strategy. In Australia, the firm is mainly a service provider and supports this service with its own leasing finance company. As a result of this approach to its business, it is in the firm's interest to develop robust machines and to recover worn and damaged components from the firm's customers for remanufacture. Establishing the Eco Manufacturing Plant at Zetland has enabled the firm to increasingly realise this objective.

A key facet of this approach to the business is an innovative technology system, Aurora, which tracks all consumables and recoverable spare parts. Developing Aurora extended the firm's component recovery rate from 80 per cent to 98 per cent. Aurora creates further efficiencies by planning a delivery trip where new components are provided to the customer at the same time as the old are recovered, thus further cutting costs.

According to Fuji Xerox managers, an important contributor to the success of the remanufacturing operation is the Engineering and Development Group who are constantly investigating ways to reuse components and materials. The Group has developed programs in electronics, lasers, mechanical subsystems, fused rollers and xerographic modules and has shown that even rectifying small defects, that once would not have been identified, can yield major results. This work has resulted in a range of new technologies that increase the scope and effectiveness of the remanufacturing process. For example, the redesign of a 15 cent spring on a roller saved the Australian company AU$1 million per annum and the US company US$ 40 million.

## Environmental commitment

Fuji Xerox's goal in this area of operation is: *"Waste free products from waste free factories"*. All processes in the factory aim to protect the environment. Not only are parts renewed or recycled but the technical processes involved in achieving this have been developed to eliminate environmentally damaging emissions, pollution and waste. For example:

- All solvents have been eliminated from the cleaning of parts and components.

- Frozen carbon dioxide (dry ice) is used under high pressure to clean components, a process that creates no liquid wastes or pollutants.

- Environmentally 'neutral' bicarbonate of soda is used under high pressure to remove the old coating from the fuser rollers used in printer/copiers. The spent bicarbonate of soda can be reused as an industrial water softener.

- The carbon by-product of waste toner is extracted and can be used as a combustion agent in steel making.

- All unusable metal parts are sent to Sims Metal to be recycled.

- Reduction in energy use is achieved through the implementation of a range of initiatives and monthly tracking to evaluate improvement through these programs.

The company is also undertaking ongoing research and development into ways of reducing all packaging waste through reuse of a range of packaging materials, including plastics. The Centre is working with Veolia Waste Management to solve these problems.

The firm also recognises, however, that there are some ongoing negative environmental aspects of the industry. The circuit board has a comparatively high environmental impact; complex parts require intensive energy systems so the total life cycle is more costly and components also include lead and sulphur.

# The eco manufacturing social system

## Leadership for change

Fuji Xerox has always had some involvement in remanufacturing, although prior to 1993 any remanufacturing was of the whole machine. The United States Xerox Corporation initiated the idea of remanufacturing components, and Dan Godamunne, now General Manager and a highly skilled engineer, was part of the group working on the initial project. Over time confidence in the product has developed, savings have become evident and remanufacturing has become a core part of Xerox Corporation's business worldwide. Worldwide Xerox has now established remanufacturing factories in nine locations in Europe, South America and the US. Fuji Xerox has four locations in Japan and the Asia Pacific region.

The previous sales model at the Fuji Xerox Australian operation was that machines were sold on leases for 3–5 years with a service contract. This was an expensive model because managing spare parts was difficult. Used parts were stripped and put on the new product assembly line. Dan Godamunne suggested eco manufacturing as a way of saving on the high cost of imports. According to one source, most managers in the firm at the time thought that the remanufacturing model could not even save two hundred dollars, but 'the firm gave him the green light because of his credentials'. Supported by the Director of Manufacturing and Supply, a long-term employee with considerable environmental commitment and political expertise in the firm, Dan Godamunne now leads a highly successful technologically based initiative.

## Human resource factors

A major issue in achieving high production targets in the new plant was to engage manufacturing and stores workers in creating a high performance operation. This was achieved through human resource initiatives, teamwork and financial reward systems. Morale is high, and staff turnover has generally been low over the last eight years, which has been a period of growth for the company.

The organisational structure is one of the main causes of the high level of innovation in the plant. It is simple, consisting of four tiers: General Manager, Operations Manager, Team Leader/Supervisor and general work force member. Staff are assigned to teams and consulted about approaches to work. The staff have given good feedback on this structure and process. Each relatively autonomous team is responsible for quality, engineering and production capacity around products or product groups, for example, xerographic modules or lasers. The product-based team structure promotes multi-skilling, enhances communication around problem identification and problem solving, builds deep expertise and cumulative experience and ensures that improved quality is constantly built into the work process. Within the limits of a manufacturing company, the company has introduced flexibility in its work practices. It allows staff to use sick leave to care for family members who are unwell and has flexible starting and finishing times around the core hours of 9.00am – 4.00pm.

The plant has also engaged in systematically building the human capabilities of its staff. Staff members are offered a range of developmental opportunities and most have had training in various aspects of people management. There are opportunities to undertake relevant professional technical courses that are supported by the company in the form of time and money. Staff can also gain some training through the Fuji Xerox intranet training system which offers courses including various management skills training courses.

Fuji Xerox also offers training to its customers. This includes developing an understanding of the technology and of the company's approach to business. Customers identify talented people in their own organisation who are then trained by Fuji Xerox to identify problems, install remanufactured components and ensure that damaged components are returned for remanufacturing.

## Cultural change

The cultural shift at Fuji Xerox Eco Manufacturing Centre has been significant. The considerable industrial unrest at the Mascot plant has eased since the move to Zetland, and a very different work culture has evolved. The nature of the work at Zetland seems to have generated a sense of worthwhile purpose. The notion that second-hand products are of higher quality is at first hard to comprehend, but both customers and the service operators now demonstrate an improved level of confidence in the remanufactured product form.

The Australian operation is considered to be the example of 'world best practice' in the process and staff take considerable pride in this achievement. The Australian company has experienced significant growth and the quantity, quality and variety of parts being remanufactured is continually expanding. On average, sixty remanufacturing programs are developed each year. All members of management agree that the process is now a core business function, fully embedded in the business and a focus for organisational identity.

Staff diversity was formerly considered a problem for the organisation. Cooperative leadership has redefined this issue and the company now benefits from the richness of its cultural mix. The company established teams in each of its key production areas such as electronics and lasers. Each team was responsible for engineering and program development. This structure led to a close working relationship between the engineers and the production workers and joint ownership of the production targets and product quality. Quality was developed 'in-the-line by-the-line'. As the teams identified problems with a product, they assumed responsibility to fix these problems. Senior managers see this team structure as central to their success in managing their people. The team organisation has led to people feeling valued and responsible. As a result individuals strive to achieve and want the company to be successful. *Strong team work*

The eco-manufacturing process builds on the quality control systems already in place throughout the company, particularly the ISO 9001 (Quality Endorsed Company), ISO 14001 (Certified Environmental Management) accreditation and the systematic processes that accompany them. The Australian plant is both ISO 9001 and ISO 14001 accredited and has in place systems and procedures to ensure that all products leaving the site receive a full quality guarantee. The ISO 14001 Enviro Accreditation is integrated into the company's Quality Management Systems and audits of ISO 9001 also include ISO 14001 Progress Audits.

Occupational health and safety (OH&S) is viewed as part of the quality process. The Director of Manufacturing and Supply in the Zetland and Mascot offices personally undertakes the site inspections for OH&S in order to demonstrate its importance to staff. Similarly, all quality processes including environmental standards have been given great importance in redefining the business. Management's position is that commitment to this standard of environmental accreditation will keep remanufacturing as the core business function. The regular audits associated with the accreditation identify the company's environmental impact and require steps to be taken to solve any identified problems.

## Influencing others

The Eco Manufacturing Centre is now involved in a number of partnership arrangements for sustainability. For example it has actively involved the academic community in joint projects with several other companies to solve sustainability problems. The firm is highly influential in the wider community through these contacts and commits considerable resources to raising community awareness through demonstrating its environmental initiatives to study groups and educational tours. These activities are listed in the Fuji Xerox Sustainability Achievements (Fuji Xerox Sustainability Report 2010).

In a range of projects conducted over the last fifteen years, Fuji Xerox has also attempted to reduce negative impacts other corporations have on the environment. These projects include partnering with research institutions to progress understanding of life cycle analysis as it can be applied in the industry, providing opportunities for schools and university teachers and lecturers to learn from study tours on site, collaborating with other industry groups and organisations through its membership of the Buy Recycled Business Alliance (BRBA) and a range of other projects.

Current initiatives are geared to the development of a 'gateway' sustainability code for all suppliers including specific environmental, social and governance requirements for high risk suppliers. Paper fact sheets have been developed to educate stakeholders on responsible paper sourcing. The company is working to develop the market for certified papers and in 2009–2010 suspended one supplier for not meeting Forest Stewardship Council Standards.

## Phases in the development of eco manufacturing

*Phase 1 - Establishment and early achievements:* The Eco-Manufacturing Centre was established by Fuji Xerox Australia in 2000 in the form of a dedicated parts-remanufacturing and recycling facility. Management at the Eco-Manufacturing Centre, firstly under Graham Cavanagh-Downs and then under Dan Godamunne, focussed on developing technological capabilities to enable remanufacturing. In the view of these managers, the success of the Centre has rested upon both technological advances and the development of a new workforce culture. In the latter regard, multi-skilled technical experts with a broad understanding of remanufacturing have been drawn from culturally diverse backgrounds and encouraged to communicate with international stakeholders, both internal and external to the firm. The Centre quickly became well known for its high performance culture associated with high workforce commitment. Innovative processes and technologies with low environmental impact were installed and developed. Supply chain and waste management relationships were established to minimise emissions of all kinds and eliminate waste. Another strategic focus has been on developing client networks for integrated systems of supply and return of used parts for repair and redesign.

Figure 3.3 sets out the history of the Centre and the shift to extended producer responsibility at Fuji Xerox. According to Dan Godamunne, 'Zetland was the first official remanufacturing plant in the Fuji Xerox world'.[8]

**Figure 3.3** Fuji Xerox achievements in extended producer responsibility

| 1990 | 2000 | 2005 | 2007 |
|---|---|---|---|
| Fuji Xerox Company Limited starts recycling parts and consumables. | Fuji Xerox Australia opens a dedicated parts remanufacturing and recycling facility in Zetland, the Eco-Manufacturing Centre, and wins a United Nations Global Environment Award. The Fuji Xerox Company Limited Asset Recovery plant in Japan achieves zero landfill. | Fuji Xerox Australia commences shipping old equipment, parts and consumables to Thailand for 99% resource recovery recycling or remanufacturing. | The Eco-Manufacturing Centre meets Fuji Xerox global manufacturing standards and expands remanufacture activity to include complex sub-assemblies for the parent. |
| **1993** | | **2006** | **2008** |
| Remanufacturing of used parts is devised and developed at Fuji Xerox Australia. | | The Eco-Manufacturing Centre becomes a total waste management centre, accepting old equipment, parts and packaging from all Australian customers for remanufacturing and recycling at Zetland or for transfer to Thailand. | Fuji Xerox Company Limited opens an integrated recycling system for used equipment and cartridges collected across China. Its disassembling and recycling capacity will cover up to 15,000 machines and 500,000 cartridges per year. |
| **1998** | **2004** | | |
| Closed loop recycling technology is adopted by Fuji Xerox Company Limited. | Fuji Xerox Company Limited opens an integrated recycling centre at Chonburi in Thailand to provide an end-of-life recycling facility for its nine operating subsidiaries across the Asia Pacific region. | | |

*Source*: Fuji Xerox Australia 2008, p. 27.

*Phase 2 - Differentiation between recycling and remanufacturing:* Building on the success of the Zetland centre, Fuji Xerox opened an integrated recycling centre in Thailand in 2004 to provide end-of-life recycling for its operations in the Asia Pacific. The aim

---

[8] Dexter Dunphy and Suzanne Benn, Interview with Dan Godamunne, General Manager Fuji Xerox Eco-Manufacturing Centre, 11 February, 2009.

was to obtain 99% resource recovery. The outcome has been that the Australian plant has increased its capacity to deliver high value-added remanufacturing, while the different skills base in Thailand has resulted in the plant there focussing the majority of its efforts on low value-added recycling. The key management strategy is that appropriate skills, capabilities and technologies need to be allocated for the differentiated processes of remanufacturing and recycling. Recent achievements at Zetland include the remanufacturing, in 2008–2009, of 250,000 parts and sub-assemblies, including many types of mechanical assemblies, complex electronic boards, electrical and optical assemblies and fusing/feeder rollers, saving AU$6 million over the cost of purchasing Xerox supplied alternate parts. The Eco-Manufacturing Centre's achievements have resulted in it being placed in the United Nations Global 500 Roll of Honour for environmental achievement in 2000 (Fuji Xerox Australia n.d.), and if is set to become the Fuji Xerox Asia Pacific hub for the remanufacture of complex sub-assemblies. Targets are set each year for the establishment of new remanufacturing programs that enable the Centre to track its key performance levels of complex sub-assembly remanufacture. The fact that the plant at Zetland is now seen as the global benchmark is prompting the shift into Phase 3.

*Phase 3 - Rolling out the new model:* The beginning of this phase was marked, early in 2009, by the appointment of a new Fuji Xerox President with a manufacturing background, Mr Tadahito Yamamoto. His aim is to drive the company beyond remanufacturing to full asset utilisation, which he believes is vital in a time of economic downturn. Everything the company has is viewed as an asset, including skilled people, hardware, location, software and intellectual property. The current challenge is total asset management. In addition, there is a need to incorporate in the company's business model recognition of future costs and other associated issues of the emerging low carbon economy, such as shipping and transfer of material resources. The Fuji Xerox President is also concerned that launching new products is very costly, and consequently he wants to extend product life from two years to five years to reduce launch costs and the environmental impact of product turnover. If and when this takes place, the aim is to ensure that product upgrades can be made remotely.

## Sustainability challenges at Fuji Xerox

The challenge of the innovation process is that many large multi-national companies regard R&D as the prerogative of their headquarters, where it is often centralised, rather than in subsidiary units that are in direct contact with various markets. The issue is one of localism versus globalism. This is an issue underpinning the challenges facing Zetland as it affirms its value in the global organisation. Nevertheless, the president has recognised the value of the Zetland Eco-Manufacturing Centre best practice model developed in Australia and is backing its diffusion to other centres. There are interesting future challenges in achieving this. One key issue is how to accommodate cultural differences in the distribution and levels of skill in other countries where manufacturing occurs. Another is how to transfer the successful but complex process of implementing transformational culture change used to build employee commitment, engagement and mult skilling.

A further sustainability challenge lies in the decision to move recycling offshore to Thailand. While such geographical differentiation admittedly adds to the products' carbon footprint, the company argues that the strategy still allows for carbon reductions through reducing virgin resource input and encouraging design for disassembly. The long-term aim is to reduce recycling to the minimum – as Dan Godamunne puts it: "Is there a need to recycle when we can remanufacture?"[9] Currently approximately 76% of items returned can be remanufactured, but the direction is to increase the percentage remanufactured through design for disassembly.

## Discussion

In its approach to remanufacturing and its people management, Fuji Xerox is attempting to become a sustainable corporation. This case study shows that the company is well advanced toward the goal of achieving sustainability, in both its policies and practices. In order to assess Fuji Xerox more comprehensively on a scale of sustainability, a developmental model, 'Phases in the Development of Corporate Sustainability', which integrates human and ecological sustainability, can be used (Dunphy, Griffiths & Benn 2007). The model identifies distinct attitudes and approaches that organisations take to achieving sustainability and ultimately a fully sustainable world.

These attitudes and approaches are represented as a series of steps. The steps are:

1. rejection
2. non-responsiveness
3. compliance
4. efficiency
5. strategic proactivity
6. the sustaining corporation.

The model makes a distinction between 'human sustainability' and 'ecological sustainability'. Human sustainability is defined as 'building human capability and skills for sustainable high level organisational performance and for community and societal wellbeing'. The ideal of ecological sustainability in a corporation is 'redesigning organisations to contribute to sustainable economic development and the protection and renewal of the biosphere'.

Although Fuji Xerox initially engaged with the sustainability concept in order to save on immediate costs, when the steps towards sustainability are considered, Fuji

*[handwritten marginal note: Developing human & Ecological Sustainability]*

[9] Dexter Dunphy and Suzanne Benn, Interview with Dan Godamunne, General Manager Fuji Xerox Eco-Manufacturing Centre, 11 February, 2009.

Xerox emerges as a company operating mainly at level 5, with some activities placing it at level 6.

# Human sustainability

## Strategic human sustainability

Flexible, team-based work practices are characteristic of strategic corporate sustainability. The high level of staff retention at Zetland allows the organisation to develop and use its people (intellectual capital) individually and in teams in order to extend the range of products the company produces. The fact that each team is a relatively autonomous unit that is responsible for its own engineering, quality and production capability is also a strategic capacity. The processes of remanufacturing are well established within the firm and this approach to the business will continue even if key individuals leave. These are enlightened human resource practices and achievements that assist the company to pursue strategic sustainability successfully.

# Ecological sustainability

## Strategic ecological sustainability

Fuji Xerox's long-term commitment to the environment is demonstrated in its identified goal of achieving waste free products from waste free factories. Some of the processes such as remanufacturing and re-engineering also give the company a competitive business advantage. As stated, currently about 76% of all parts used in Australia are remanufactured; the target is to remanufacture 100% of parts. The forwarding of waste products such as carbon and bicarbonate of soda from Fuji Xerox to other companies results in the conversion of what was formerly regarded as waste into useful commodities. Fuji Xerox also makes strategic use of the principles of industrial ecology through developing relationships with E-Sims/Sims Metal, who recycle waste metal, and Veolia who pride themselves on environmental waste management. These strategic partnerships advance both ecological and human sustainability.

# Conclusion

Fuji Xerox Australia takes justifiable pride in its technological achievements in relation to re-engineering and remanufacturing. It is also proud of the financial contribution these approaches have made to the business and the benefits to society of its approach to the environment and corporate culture. The company recognises the contribution of its people and their importance in the company's success. Fuji Xerox Australia has championed a culture which has fostered innovation and the growth of intellectual capital, demonstrating that, in this case, what is good for people and the environment is also good for business. Remanufactured products help protect the environment by reducing landfill and utilising valuable existing resources. Remanufacturing also conserves the energy and materials that would otherwise be used to make new products. The principles behind this work are applicable to a wide range of other products such as automotive systems, computers and electrical equipment, cameras and photographic equipment and household electrical goods. These principles need to be diffused across a wider range of industries to reap the potential benefits for the Australian economy, society and the environment.

# References

Fuji Xerox Australia n.d., 'Eco Manufacturing: FXA now exports remanufactured parts and components to the Asia Pacific region', <www.fujixerox.com.au/about/eco_ manufacturing.jsp>, accessed 4 April 2009.

Fuji Xerox Australia Sustainability Report 2008, <http://www.fujixerox.com.au/ docs/fxa_ sustainability_report_2008.pdf>, accessed 13 November 2010.

Fuji Xerox Sustainability Report 2010, <http://www.fxasustainability.com.au/> accessed 21 October 2010.

Benn, S & Dunphy, D 2004, A Case of Strategic Sustainability: the Fuji Xerox Eco Manufacturing Centre. *Innovation: Management, Policy and Practice*, 6, 258-268.

Dunphy D, Griffiths A. & Benn S 2007, *Organisational Change for Corporate Sustainability*, 2nd edn, Routledge: London.

# Case 4

# Hewlett Packard's supply chain

JOHN CHELLIAH AND SUZANNE BENN

## Introduction

Hewlett Packard (HP) is a technology solutions provider to consumers, businesses and institutions globally. It has products and services that span IT infrastructure, personal computing and access devices, global services and imaging, as well as printing for consumers, enterprises and small and medium businesses. This case focuses on HP as a global company and associated global supply chain issues.

## Method

Information for this case study was obtained from interviews conducted by the authors with the Annukka Dickens, Environmental Manager South Pacific, Hewlett-Packard Australia and Michael Wagner, Solution Architect, Hewlett-Packard Australia, from secondary documents and from company websites.

## Sustainability strategies at HP

For this global organisation, the concept of sustainability is certainly not new. HP's commitment in upholding corporate citizenship and sustainability practices has a long history in the organisation (Lowitt & Grimsley 2009). In implementing sustainability, HP management has attempted to go beyond product design and technical issues to infuse sustainability principles across the organisation. HP's network of environmental professionals – ranging from front-end staff who deal directly with stakeholders, to packaging engineers, to HP suppliers – all play a part in ensuring that the sustainability initiatives both perpetuate and build on the financial strengths of the organisation (Sairanen 2006).

HP's sustainability strategies can be identified as follows:

# Strategy 1: Maximise market leader position

HP is the leading company in its industry. It recognises that going beyond compliance to actively contributing to government rule-making, as well as taking on pro-active strategies to pre-empt policy, enhances their reputation and enables the firm to take leadership to increase competitiveness. An example of this is the expansion of HP's Planet Partners return and recycling program, which offers authorised retail recycling locations for HP inkjet and LaserJet print cartridges nationwide. The first authorised retailer to pilot the program is Staples, and HP plans to introduce additional retail partners in the near future. HP kicked off the pilot program with Staples in conjunction with America Recycles Week, offering customers a free and convenient way to responsibly recycle HP cartridges. Many customers already bring their used cartridges to retail stores when purchasing replacement cartridges, and in-store recycling options provide an added customer convenience. Customers can drop off their used HP cartridges at more than 1,500 Staples locations across throughout the United States for recycling through HP Planet Partners (Planet Ark website 2009; HP Website 2008).

For their commercial customers, HP provides a take-back program called Asset Recovery Services where customers return their products to HP for responsible recycling. Returned products that still have retained value are refurbished and resold on the client's behalf. In some instances, the recouped value can be returned to the customer allowing them to offset some of the costs when they buy a new product from HP.

HP also offers a remanufacturing solution, the HP Renew Program, on many of its hardware products (HP website 2010c). Products or parts that are returned are remanufactured and certified to be good as new. They are then given the same-as-new warranty status.

# Strategy 2: Building collaborative relationships

HP is a signatory to the United Nations Global Compact, which is a voluntary initiative relating to human rights, labour, the environment and anti-corruption. The initiative provides business opportunities through more efficient management of resources as well as a key proof-point that demonstrates environmental leadership to those customers whose procurement decisions take the environmental sustainability of suppliers into account. It also provides leverage when seeking to influence other organisations to adopt the sustainable initiatives for their own systems.

HP approved suppliers are treated as partners and given support in the form of tools and expertise in order to meet sustainability requirements. However, the firm maintains standards by constant vetting of suppliers, and is willing to terminate contracts if these standards are not met (Lowitt & Grimsley 2009). There is also an Environmental Advisory Council that has members including non-government organisations. It lends not only credibility but new sources of knowledge and alternative viewpoints on sustainability issues and challenges.

## Strategy 3: Turning environmental priorities into competitive advantage

One facet of HP's competitive advantage comes from ensuring that recycling of electronic goods (one of HP's environmental priorities) is embedded in the way they do business. Instead of looking only at the disposal phase of a product, HP implements environmental considerations throughout the life cycle of each product through a competitive core design strategy. However, HP is mindful of the fact that recycling as a cost in the supply chain needs to be minimised, and it has implemented the HP Renew remanufacturing program described above. As HP's Environmental Manager South Pacific points out:

> ...instead of having twelve screws in your printer why don't [we] just eliminate all the screws and have snap-on features? [Snap-on features are fixtures that affix product materials and components together similar to screws]. Using snap-on features mean reducing the time to separate plastics...thus saving] money in the recycling proportion and having a competitive advantage over someone who doesn't do that. (Sairanen 2006)

Sustainable and innovative product design ideas have led to an approach by HP which focuses on reducing materials used in the making, packaging and delivery of the product. By minimising the amount of packaging that surrounds each product, HP effectively minimises the space that the products take up, hence reducing transportations costs that come with airfreight or sea freight methods. In 2005, for example, the number of PCs that could fit on one shipping pallet rose from 28 to 40 units, thus effecting a 40% reduction in the energy required for shipping (HP 2006b). HP has expanded the use of plastic pallets, which are 70% lighter than wooden ones, saving fuel in transport.

HP Product designers are trained on the company's 'Design for Environment' principles. HP has its own recycling centres in Europe and North America and first-hand learning from these organisations has translated into a sustainable perspective across the organisation. Management of the recycling process has helped the company to learn from its experiences in product design and recycling and to translate these experiences into effective, better approaches to product disassembly, recycling and remanufacturing. Subsequently, this learning has been incorporated into the training of product design engineers. During the design process, there is opportunity to assess an element and remove it from the product in order to enhance the product's ability to be recycled. Designers are also trained in the regulations around the world that require HP to remove certain hazardous components (e.g. batteries) from the products before they recycled.

## Greening the supply chain

Green supply chain management involves the integration of both environmental and supply chain management in order to reduce a company's impact on the environment while improving business performance. As a first mover in the industry, HP has had to re-educate their partners and suppliers along the way in fulfilling their environmental corporate objectives. As awareness of environmental sustainability builds in the IT industry and overall business environment, HP finds itself in an interesting position where customers are now demanding solutions for the dilemmas they face as a result of global environmental regulations. Effectively, the market and regulatory forces have provided the impetus for the industry to catch up with HP. This push from the supply chain provides HP with more support for their sustainability initiatives.

The Sustainable Product Service System (SPSS) is an example of HP's continued investment in creating innovative services and product systems, which are all part of harnessing their competitive advantage. HP is a leader in the industry in Australia and globally in this holistic approach to product development (Hargroves *et al.* 2007). The SPSS is an aspect of a total cost of ownership (TCO) approach founded by HP and Gartner in the 1990s. The SPSS represents a 'cradle-to-cradle' approach, where people are offered solutions as products. Need-focused solutions are inherently more sustainable than products as they offer the value of use rather than the product itself (Tukker & Tischner 2006). Customers are provided with a solution which incorporates the hardware, software and services in a Sustainable Product Service System (SPSS).

**Figure 4.1** Total cost of ownership

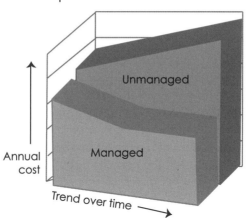

*Source*: HP TCO 2006.

Total cost of ownership (TCO) is defined as the cost of procuring, deploying, managing and maintaining Information Technology (IT) systems (HP TCO 2006). Figure 4.1 is a graphical representation of how a much lower total cost of ownership can be achieved over time by using a managed approach for all the costs involved in the life cycle of IT systems. This is where the SPSS fits in using a life cycle approach. The 'unmanaged' approach results from purchasing decisions made only on initial hardware cost without considering the impact for on-going support and services costs. The reality is that initial hardware costs of the technology amounted to only 20–25% of the total cost, whilst post-deployment costs may constitute up to 80% of total IT expenses (Green Growth Website 2007; Wagner 2006).

Using the approach depicted in Figure 4.2, HP is able to measure TCO cost savings for customers in the management, maintenance, upgrade, and support of their overall IT environment. The conception of SPSS was customer driven. Customer feedback signalled demand for a sustainable model of funding which allowed customers to run and manage their IT system in a responsible and efficient manner (Preston 2001).

**Figure 4.2** Life cycle approach to TCO reduction

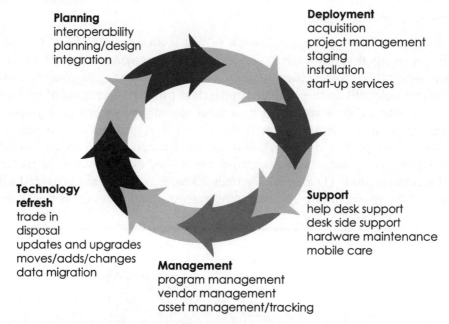

**Planning**
interoperability
planning/design
integration

**Deployment**
acquisition
project management
staging
installation
start-up services

**Technology refresh**
trade in
disposal
updates and upgrades
moves/adds/changes
data migration

**Management**
program management
vendor management
asset management/tracking

**Support**
help desk support
desk side support
hardware maintenance
mobile care

*Source*: HP TCO 2006.

Despite being initially customer-driven, when it comes to selling the system, HP has experienced a low level of market penetration, although the market is now maturing (Hargroves *et al.* 2007). HP has tried to get around this hurdle by training its executives and sales team to sell the benefits of the SPSS, such as:

- economic flexibility and high return on investment
- better service performance and flexibility with less risk
- lower costs for product end-of-life management.

## Conclusion

As the SPSS example highlighted, until there is legislative force behind sustainability initiatives, 'short termism' on the part of customers (and investors) may prevail. This may in turn inhibit highly innovative product development that could address some of the major toxic waste issues currently caused by the IT industry. HP clearly sees sustainability in business advantage terms: reducing waste, gaining a competitive edge, increasing employee engagement and stimulating innovative practices. Yet sustainability in the business context is a delicate balancing act and the company makes careful decisions about where it places its priorities. Sustainability at HP is very much dictated by the firm's global agenda. This case highlights the fact that sustainability needs to be integrated into all decision making within the organisation, its customers and the supply chain for long-term business, environmental and social benefits to be achieved. In particular, it highlights the complexity of supply chain issues and the need to consider the institutional drivers that prompt or inhibit corporate responses to sustainability.

# References

Green Growth Website 2007, viewed 19 August 2010, <http://www.greengrowth.org/download/greenbusinesspub/Greening_of_the_Business/Civil_Society/David_Ness_ Sustainable_Product_Service.pdf.>.

Hargroves, K., Stasinopoulos, P. & Smith, M. 2007, 'Sustainable IT' through 'Sustainable Product Service Systems', a case study of Hewlett Packard, The Natural Edge Project (TNEP), Australia.

HP Website 2010a, viewed 26 August 2010, <http://www.hp.com/hpinfo/globalcitizenship/enviro/energy/transport.html>.

HP Website 2010b, viewed 26 August 2010, <http://www.hp.com/hpinfo/globalcitizenship/enviro/design/packaging.html>.

HP website 2010c, HP Renew, viewed 24 November 2010, <http://www.hp.com/united-states/renew/why_buy_refurbished.html>.

HP Website 2008, viewed 26 August 2010, <http://www.hp.com/hpinfo/newsroom/press_kits/2008/retailrecycling/FAQ_Planet_Partners.pdf?jumpid=reg_R1002_USEN (PDF)>.

HP Website 2006a, viewed 21 September 2006, <http://hp.com.au/recycle/supplies/?jumpid=reg_R1002_AUEN>.

HP Website 2006b, 'New South Wales and HP Commit to Sustainability', viewed 21 September 2006, <http://search.hp.com/gwaueng/query.html?sh=1&charset=UTF-8&la=en&qt=New%20South%20Wales%20and%20HP%20Commit>.

Lowitt, W & Grimsley, J 2009, 'Hewlett-Packard; Sustainability as a Competitive Advantage, Accenture Institute for High Performance Case Study', viewed 24 November 2010, <http://www.hp.com/hpinfo/globalcitizenship/environment/commitment/accenturestudy.pdf>.

Planet Ark Website 2009, viewed 26 August 2010, <http://cartridges.planetark.org/about/manufacturers.cfm>.

Preston, L 2001, 'Sustainability at Hewlett-Packard: From Theory to Practice', *California Management Review*, vol. 43, no. 3, p.34.

UN Website 2010, viewed 27 August 2010, <http://www.unglobalcompact.org/aboutthegc/thetenprinciples/index.html>.

Sairanen, A (Environmental Manager South Pacific), Hewlett-Packard Australia, interview with authors on 12 May 2006, Sydney.

Tukker A & Tischner U 2006, 'New Business for Old Europe Product-Service Development', *Competitiveness and Sustainability*, Greenleaf Publishing, Sheffield.

Wagner, M (Solution Architect), Hewlett-Packard Australia, interview with authors on 1 September 2006, Sydney.

# Case 5

# IKEA: A company's progression to a strategic approach

BRUCE PERROTT AND DEXTER DUNPHY

## Introduction

The IKEA vision is to create a better
everyday life for the many people.
We make this possible by offering a
wide range of well-designed,
functional home furnishing products
at prices so low that as many
people as possible will be able to
afford them.

In this case we document the increasing importance given to sustainability initiatives by international retailer IKEA over its 60 year history. The case describes a changing emphasis from treating sustainability as primarily about eliminating waste to giving sustainability an expanded role in the company's core business strategies.

This case study was written from information collected from face-to-face interviews and numerous internal documents to which access was granted. The documents utilised include speeches, surveys, reports and plans. A total of seven face-to-face interviews were conducted by the authors with IKEA personnel from various international locations and positions. Interviews were audio recorded in addition to notes being taken by the interviewing authors during the period April to August 2010. Appendix 5.1 lists interviewee's names and positions.

# Background to IKEA and sustainability

IKEA's founder, Ingvar Kamprad, began his business in the 1940s, selling pencils and matches. The initial market was in Smaland in Southern Sweden where individual sales calls were made and orders delivered using a milk cart. From this simple beginning he developed a mission to produce well designed functional furniture at the lowest price possible without compromising the product's quality. This concept proved popular and the IKEA catalogue was made available in 1951 and the first retail store opened in Almhult, Sweden in 1958. After that, a new IKEA store was opened in Scandinavia every few years as consumers warmed to this novel way of shopping for their home furnishings. IKEA eventually went international and the first Australian store opened at Artarmon in 1975. In 2010, there were 300 stores worldwide in 35 countries. In 2009, 660 million shoppers made visits to these stores that together recorded € 22.7 billion (AUD 32.22 bn) in turnover. The annual catalogue is a mainstay of the IKEA marketing appeal and method of communication to potential customers. Each year 190 million catalogues are printed in 25 languages and made available in 35 countries.

IKEA's distinctive value proposition for customers is that they will offer 'a wide range of well-designed functional home furnishings products at prices so low that as many people as possible can afford them'. In an early initiative, IKEA pioneered the 'flatpack' concept which allowed unassembled furniture to be transported economically and then assembled by the customer in their own home. This created significant efficiencies in production, transport and storage costs; it lowered prices for consumers and benefited the environment.

IKEA's awareness of the company's environmental responsibilities began in 1986 as a consequence of the formaldehyde crisis in Denmark, when the health implications of this chemical became known. Shortly after this event, IKEA appointed an environmental manager named Russel Johnson. IKEA's first environmental policy was produced in 1991, with a detailed action plan following in 1992. In 1993, IKEA completely banned Formaldehyde in the lacquer used to coat its products. This was also the year that the IKEA catalogue became the first totally chlorine free publication as a result of close cooperation between IKEA and Greenpeace.

By 1999 IKEA was extending its concerns to social issues. They appointed a 'children's ombudsman' to look after and into the interests of children around the world, particularly in less developed countries. One person was employed in India to work on social projects and training; projects included work on the construction of a school and supporting teacher training. This was the beginning of the company's ongoing concern for the interests of the world's children, and IKEA are now globally the biggest corporate donors to both UNICEF and Save the Children.

In 2000 IKEA launched a code of conduct called 'IWAY' which is the company standard when purchasing home furnishing products. This code laid down minimum requirements for all IKEA suppliers regarding social and work conditions for their employees, the environmental impact of their operations and the use and sourcing of raw materials. In 2001 the company also worked with the World

Wildlife Fund (WWF) to begin a project to support the management of sustainable forests.

In 2002, IKEA launched an e-learning program on their intranet that addressed various social and environmental issues. The first Social and Environmental Responsibility Report was launched on the Web in 2004. The same year IKEA joined the Global Compact which is a United Nations initiative based on key principles, which include the areas of human rights, labour, environment and anti-corruption.

In 2006, the IKEA social and environmental strategy was launched with a particular focus on leadership and competence, buildings and infrastructures, community involvement, energy and transport. This was also the year the IKEA Social Initiative was formed for the purpose of handling global donations for social projects within the IKEA group, the Inter IKEA Group and the IKANO Group. The same year, IKEA implemented the EEU directive on waste electrical and electronic equipment (WEEE). As a result each store now recycles electrical and electronic equipment. Customers can also return potentially hazardous waste resulting from products sold such as spent batteries and low energy bulbs.

IKEA started its IKEA Goes Renewable Project (IGR) in 2006. Under this project policy, IKEA stores, warehouses, offices and Swedwood units (IKEA owned suppliers) must be supplied with 100% renewable energy and undertake to progressively improve energy efficiency.

IKEA and WWF started several joint projects in 2007 aimed at reducing emissions caused by the operation of the business. IKEA also launched a new global strategy known as 'Meeting the IKEA Way'. This was about making it easier for all co-workers to meet and travel in the most effective and efficient way. The slogan was 'meet more, travel less'. At this time, the company also decided to only serve and sell UTZ certified coffee that was traceable all the way back to sustainable plantations. UTZ coffee is grown with consideration and care for social communities and the local environment (UTZ website).

In 2008 IKEA held the first of several futures research workshops with the theme 'IKEA 2020: Leading the Way to a Sustainable World (together we will create a vision for a sustainable business in 2020). Seventy five internal and external stakeholders gave their views on how to handle long-term challenges and business opportunities. The input from these forums was used as a basis for developing future sustainability strategies for the company. Part of the IKEA secret to success has been that all its products are designed in-house. There is a strong emphasis on innovative design and now sustainability principles are incorporated into the design phase.

A new IKEA platform for sustainability was launched in 2009 with the theme of the 'never ending job'. Co-workers and customers were involved to produce a first edition of 'Never Ending List' which outlines the many ongoing sustainability issues to be addressed by IKEA. 2011 will be the first year that IKEA Australia produces its own separate sustainability plan which reflects these issues.

At the launch of the Never Ending List in Australia in March 2010, David Hood, Country Manager, IKEA Australia, said: "IKEA recognises that until this point we have potentially been part of the (environmental) problem. The Never Ending List is our commitment to contribute to the solution. This means that we will constantly review what we are doing and how we are doing it to ensure we have as little impact on the social and natural environment as possible".

## Sustainability today in IKEA

A strong commitment to ongoing sustainable management is expressed in the 2009 Sustainability Report's introductory letter written by the Thomas Bergmar, Sustainability Manager, IKEA Group:

> ....I am particularly pleased with our strategy work for 2010–2015. Each and every one of our business strategies - whether local, national or global - must now clearly and systematically integrate sustainability as a part of everyday operations. I see this as a milestone. It will help us move forward in our relentless work with the many challenges we have to solve on our route to becoming a sustainable company.

The IKEA vision is to 'create a better everyday life for the many people'. This includes a lot more than just providing a great home furnishing offer. It is also about taking social and environmental responsibility in relation to IKEA customers, co-workers and the people who produce IKEA products. In particular IKEA wants to

- offer solutions and know-how that help customers live a more sustainable life at home

- use natural resources in a sustainable manner within the entire value chain

- minimise the carbon footprint from all IKEA related operations

- take social responsibility and act as a good global and local citizen

- be transparent to all stakeholders and communicate more to customers and co-workers.

In this vision statement, IKEA recognises that making improvements for a better and a more sustainable company is a never ending job because there will always be more work to be done in progressing sustainability. Actions taken and results in key areas of endeavour are documented and published in the annual sustainability report. In addition, the Never Ending List has new sustainability initiatives added which are published on www.ikea.com.au. In this way, IKEA tries to be transparent in its intent and actions and to make the company publically accountable.

IKEA is also committed to involving its stakeholders in an ongoing dialogue about the future. As an input to 2010–2015 strategic directional planning, a three day 'Future Search' workshop was held with selected internal and external stakeholders. Working together, this group defined what sustainability means to IKEA:

> Wherever we are, we act with respect to exert a positive impact on people and on the limited resources of our planet to ensure long-term profitability.

This statement summarises the IKEA commitment to sustainability management and is integrated into every one of its business strategies including the product range. One recent initiative is to find new ways for customers to take care of IKEA products at the end of their life cycle by replacing parts or returning products for recycling. IKEA has made an ongoing commitment to include its stakeholders in future sustainability planning. Key stakeholder group opinion and feedback is collected on a systematic basis along the following lines:

*IKEA suppliers and their employees:* The close relationship build up between IKEA workers and their many suppliers opens up dialogue and frank discussion about what is involved in solving the many social and environmental issues that emerge. The code of conduct for suppliers is laid down in the IWAY program, which specifies minimum requirements and what suppliers can expect in return from IKEA. There has been a series of continuous improvements since IWAY was introduced in 2000. Every three years IKEA conducts an anonymous survey to gain honest feedback about supplier relationships and to identify further issues that need to be addressed.

Of the many IKEA sustainability related initiatives, IWAY is probably the most significant in terms of its end product impact. This program was introduced in 2000, is continually upgraded and specifies minimum requirements for all suppliers of IKEA products and services. A revised schedule was introduced in January 2009. Globally, IKEA has 1,220 home furnishing suppliers, 97 global food suppliers, 18 catalogue suppliers, and 278 transport service providers. IKEA has a goal to build long-term sustainable relationships with suppliers who share similar values. Some of the critical requirements include the prevention of child labour, setting environmental standards and goals, elimination of corruption, and achieving at least a minimum standard for healthy employee working conditions. Specific IWAY checklist items include:

- start-up requirements
- general conditions
- environmental requirements
- use of chemicals
- hazardous and non-hazardous waste
- fire prevention
- worker health and safety
- housing facilities
- wages, benefits and working hours
- child labour requirements
- forced and bonded labour
- discrimination

- freedom of association

- harassment, abuse and disciplinary actions.

Regular random supplier audits are conducted both by IKEA inspectors and third-party KPMG consultants to ensure minimum conditions are maintained. As a result, while IKEA used to have around 2000 suppliers, this has now been reduced to 1400.

*IKEA co-workers:* IKEA employees are referred to as co-workers. An annual company-wide anonymous survey is conducted to gain insights into the issues that concern co-workers. In the 2009 survey, 87% of co-workers ranked as 'favourable' the statement: 'It is my responsibility to contribute to IKEA's social and environmental work'. This was up from 79% in the 2008 survey. In the 2009 survey, 79% ranked as 'favourable' the statement, 'IKEA is a company that shows in action that it takes social and environmental responsibility'. Clearly this reputation of being a socially environmental company is helpful in attracting employees who care about the environment. For example, Josja Van Der Maas, Deputy Country Manager, IKEA Australia, spoke of completing an MSc degree in Business Economics and Human Resources at the Erasmus University of Rotterdam, The Netherlands. The course included a case on IKEA and Josja said:

> As a result of that case, I fell in love with the company. Even 14 years ago IKEA was a responsible company. I joined because this is a place where I can just be myself I come from a very socially and environmentally aware family and I saw this as an opportunity to contribute.

*IKEA customers*: Each year a market survey is carried out through 'Brand Capital'. In addition, each local store conducts its own customer satisfaction survey to better understand consumer issues that affect opinion at the point of purchase.

In one environmental study IKEA conducted with 1,246 Australians aged between 18 and 64 years in January 2010, seven out of ten said that they consider environmental friendliness of a product at least occasionally when making buying decisions, while 28% stated that they rarely or never consider it.

To the question:

> Now thinking about the environment and Australia as a whole – including governments, companies and individual Australians, in your opinion, how important is it that Australia takes steps to help protect the environment?

Seventy four per cent said it was either extremely or very important.

*Communities:* IKEA participates in selective community projects in all of the countries in which it operates. Children's wellbeing and education has been a major focus of recent efforts with more than 100 million children expected to benefit from the IKEA initiatives.

*NGO's and other stakeholders:* Through cooperation and regular liaison, IKEA seeks to share knowledge that benefits relationships with various community groups such as Global Compact, Greenpeace, WWF, Better Cotton Initiative, Rainforest Alliance, Save the Children, Unicef and the ILO.

IKEA has a wide international coverage in its retail operation. Product purchases in terms of sales volume by region are: Europe 67%, Asia 30% and North America 3%. The main countries of purchase for IKEA products are;

| | | | |
|---|---|---|---|
| China | 20% | Germany | 6% |
| Poland | 18% | Sweden | 5% |
| Italy | 8% | Other countries | 43% |

What IKEA achieves by its sustainability initiatives therefore has a wide impact throughout the world.

## Staff matters

Another major focus for IKEA is the development of a sustainable work force. In 2009 IKEA had 123,000 co-workers spread across the range of working tasks:

| | |
|---|---|
| Purchasing, distribution, wholesale | 13,800 |
| Swedwood Group | 15,000 |
| Retail | 94,200 |

Numbers of co-workers per region were: Europe 99,700, North America 15,500 and Asia and Australia 7,800. Consequently IKEA's sustainability initiatives directly affect the attitudes and actions of a large number of employees. Staff turnover is one important measure of whether IKEA's own workforce policies are having their intended effect. In recent times, staff turnover in Australia has dropped from 55.3% in 2007, to 47.7% in 2008 and 31.9% in 2009. Currently, 55% of workers stay with IKEA for three years or longer. Among the improvement programs in this area, a more clearly focused recruitment strategy has resulted in the hiring of workers with appropriate competencies and who are seen as suited to working in the IKEA culture. Ongoing staff programs also include improving working conditions, paid maternity/paternity leave provisions, promoting equal opportunity, strengthening health and safety practices, providing redundancy support, undertaking sustainability training and improving communications.

## Sustainability in Australia

"We want to be judged as a change catalyst for sustainability in Australia"

(David Hood, Country Manager, IKEA Australia).

As Country Manager of IKEA Australia, David Hood first became aware of sustainability issues while being involved in the design of a recycling system for the third London store in 1996. At this time such initiatives were part of IKEA's social responsibility movement. Hood said:

> *However, in the early days we saw it in the context of how to better get rid of the waste. This was a huge cost that we had to address. Sustainability wasn't really a word. Since the late 1990's, the Group has grown hugely and has introduced numerous actions aimed to progressively improve how the*

*Company and its stakeholders interact with the environment. Sustainability has become one of our main stream responsibility areas. New initiatives are progressively introduced as we see the opportunities. For example we have reduced the number of Company cars in Australia from 24 to only 12 and these are all hybrids.*

At present there are company run stores on the East Coast and a franchise operation in South Australia and Western Australia. There are about 1,200 co-workers in the Australian IKEA operation. Sustainability planning for new stores at Tempe and Springvale comes under the management of Annie Chandler and Chris Hamilton. Chris is the IKEA National Facilities Manager. Annie began her career with IKEA as a university student working part time in Sales. She has worked through a number of operational roles in IKEA before being appointed in 2009 to her present position of IKEA Social and Environmental Manager for Australia. Annie suggests that we could have more government agency support for waste management and recycling programs and that the Western Australian government agencies are more supportive of sustainability suggestions and proposals than the Eastern states.

Batteries and light bulbs are an important part of the IKEA product range. As one of its many and ongoing initiatives Chandler says: "The Company is about to relaunch its recycling program that will ensure more customers return old batteries and bulbs to an IKEA store when they purchase a replacement product".

Richard Ansell is the Property and Expansion Manager for IKEA Australia. He said:

*My role is to conduct a market analysis in order to determine how many stores we should open in Australia, decide location, acquire and construct the approved stores ready for opening. Hence I have responsibility for facilities and sustainability.*

Annie Chandler and Chris Hamilton report to Richard Ansell. Richard summarises how he sees the approach to sustainability in IKEA:

*We take the international IKEA strategy and direction for sustainability and then produce a local business plan/strategy for five years and then detail the year ahead. We set national goals and then break them down into store by store goals. Before three years ago, I thought that sustainability was disconnected in IKEA. The high level principles sometimes did not filter down to the store level. Sustainability now has a much firmer basis and commitment.*

Globally, the company is continually working on their Never Ending List of sustainability matters to address, which include milestones that must be achieved in areas such as energy emissions, packaging, recycling targets and waste reduction. Co-workers are asked their opinions in staff surveys and progressively educated about the ways to reduce waste and save energy both at work and at home. A 2009 survey of co-workers showed that 87% thought it their responsibility to play a part in promoting sustainability in IKEA. The company sees sustainability as an integral part of the key areas of strategic focus: customers, finance and people. The main execution of sustainability strategy is at the store level where the customer comes into contact with staff, products and services offered by IKEA.

Even information about how customers travel to an IKEA store can provide actionable sustainability initiatives. For example at Logan in Queensland, one company survey found that only 1% of customers travelled to shop at IKEA by public transport; whereas at IKEA Richmond, Melbourne, the figure was 15%. A study of the public transport service found that there was an opportunity to discuss a more direct and convenient service for IKEA customers which could encourage them to use public transport.

In IKEA there is considerable attention paid to reducing packaging which is usually a substantial and integral part of transportation and storing furniture. On the Never Ending List are action plans to reduce packaging and waste and to improve recycling performance. For example, Visy have been recently appointed as IKEA's national recycling partner. This partnership has the objective of achieving 90% recycling of all IKEA waste by 2015 (now at 55% levels). The new arrangement is expected to have zero net cost to IKEA. An example of one by-product opportunity has also been shown with the development of OPTILEDGE which is a plastic corner that fits under a product pallet and allows it to be stored and moved. This product reduces the need for wooden pallets which can now be replaced by paper pallets. This innovative product is now sold to other companies.

## What the future holds

> The COP15 summit probably has been a failure, but the uncertain scenario today doesn't have any space for 'second thoughts'. Rather the opposite; it gives even more urgency to the need to keep the rise of the temperature of the planet under two degrees Celsius, to avoid that natural resources are exploited above the capacity of the earth to regenerate them and to secure that the many people have a place to live a good everyday life. (Stefano Brown, Global Sustainability Manager, IKEA)

Richard Ansell suggests that:

> In the past the IKEA style of operation with its low cost philosophy has enabled the company to implement sustainability measures with a view to efficiency gains rather than only with a focus on emissions. However, with the new more strategic approach we now take both factors into consideration.

The proposed new Tempe store in Sydney is scheduled to open at the end of 2011 and represents an investment of over AUD200 million. Specifically regarding the new Tempe store, Richard Ansell said:

> Unfortunately Tempe has come quite early in the process (i.e. IKEA's new sustainability planning process). Hence we haven't had the chance to build in all the things (i.e. sustainability actions) we would have liked. However, Tempe will have more sustainability features than the Logan store we built in Brisbane.

Josja Van Der Maas remarked that the Logan store was an advance on previous stores in terms of sustainability features built into it and agreed that "The Tempe store will be a bit more advanced than Logan but not a lot more". She mentioned restrictions imposed by the site which is an old waste dump composed of landfill and is also directly under the major flight paths for aircraft approaching and leaving Sydney airport. The underground waste means that there are limitations on the extent to which ground soil can be disturbed and the noise from flights limit use of natural airflows to cool the building.

However, all possible sustainability measures are being investigated. There will be, for example, massive water tanks for rainfall collection (1.4 million litre capacity), solar hot water systems and an onsite recycling station. The roof to be installed will have far better insulation qualities than is usual in Australia and good acoustic properties to reduce aircraft noise. The grounds will be extensively landscaped and have three times the number of trees that are normal for similar retail centres. All possible measures are being considered and where feasibility studies reveal a payback period on investment within eight years, the measures will be adopted. This criterion was confirmed in the interview with Richard Ansell.

In discussing forward planning, Richard Ansell said:

> We look at key areas of opportunity including the reduction of waste, the saving of energy, and the education of staff and customers regarding sustainability standards and practice. _We look at justifying a sustainability investment by the requirement of having that investment show a payback within eight years._ In Australia this principle is more difficult in practice as the cost of energy here is so low when compared to other countries.

Energy is definitely one area in which this criterion creates difficulties in achieving IKEA's global guidelines. IKEA Global has a policy of achieving 100% use of renewable energy in its stores. Achieving a return on investment in solar heating of hot water presents no problems, but in Australia, the price of conventional power produced by coal powered stations is so low, that the eight year payback period is probably not achievable for electricity production from renewable sources. (Australian power prices are about one tenth of those in Europe). IKEA Australia is currently researching how close they can come to reaching the company's international targets given rising energy costs here.

The social environment of the new Tempe store is also being considered in the planning process. The store is in the Marrickville Council area which is a very multi-cultural area with for example, Greeks, Macedonians, Chinese and Korean communities. The school populations are therefore very diverse. Early studies are trying to discover what positive impacts IKEA could make in the area, for example, by strengthening local employment opportunities for a diverse workforce. IKEA would like to have a positive influence in relevant social and environmental areas. Staff are already talking to the Council and community groups and listening to their concerns. It has been decided that the workforce of about 400 will be recruited both locally and from elsewhere (there is an excellent rail service which will allow staff from outside the area to commute to the new store by public transport). There will also be job possibilities for people with disabilities – as promoted in all IKEA Stores.

The IKEA approach to forward planning for sustainability in Australia begins with a review of the four corporate pillars of the company's sustainability strategy which are detailed below. In the strategic planning process, the opportunities within each pillar are discussed in terms of their potential for introduction in the overall

Australian operation and then for their potential application at the individual store level. Areas for action in 2011 include the following:

## Integration in the product range

- implement product specific communication across all locations
- update in-store communication package as per the IKEA of Sweden updates
- re-launch take-back program for batteries and bulbs
- launch take-back service for co-workers ?
- ensure that Never Ending List communication is included in all 'Store in Shape as New' and new projects
- secure a green thread in our commercial activities.

## CO2 emissions and use of raw materials

- develop a sustainable building model encompassing passive, active and renewable new projects
- measure and improve store environmental performance with a recognised tool
- review and improve the design of our new stores with a recognised tool
- retro-fit existing stores with new technologies and maximise the potential of our building management systems
- secure waste sorting at the source in all areas
- reduce CO2 emissions from our customers and co-workers travelling to our stores.

## IKEA social initiative

- establish a community engagement plan for projects
- establish a community engagement plan for each market
- identify and promote one national charity through the IKEA family
- improve co-worker awareness of IKEA social initiative activities
- roll-out IKEA charity policy implementation guidelines (global).

## Inform more people

- expand 'Our Responsibility' on IKEA.com.au
- prepare bi-monthly update on sustainability activities
- create and release an Australian sustainability report

- share best practice on a local and international level

- increase our media presence in corporate/sustainability media.

Josja Van Der Maas commented that IKEA's culture and values give a huge advantage in ensuring that these ideas are put into practice. She stressed that IKEA still has many of the characteristics of a family company and has always encouraged bottom-up initiatives by co-workers. There are also active networks within the global company where information is shared. There is still a feeling of being small while the company is in fact now quite large. The company encourages people to pursue their enthusiasms and provides them with support. In addition, every store has its own business plan and managers have key performance indicators relating to four key areas, one of which is sustainability.

There is a growing realisation within IKEA that sustainable strategies also create marketing opportunities. In the words of Richard Ansell, "We now realise that there are also some public relations and marketing gains to be made under the sustainability heading".

In addition to opportunities identified in the annual sustainability plan, there is continual dialogue regarding possible areas that can be addressed to reduce waste, conserve energy and reduce CO2 emissions. Annie Chandler says she often comes up with sustainability ideas and strategies, and then discusses their cost and feasibility with the Facilities Manager, Chris Hamilton. She commented: "It is a good partnership that helps ensure that sustainability thinking is connected to ongoing action and opportunity creation".

## New retail stores in Australia

"The ambition of IKEA is to have sustainability principles embedded in the business and to turn what is traditionally considered as an issue into an opportunity" (Stefano Brown, Global Sustainability Manager).

IKEA is gearing up for the 2011 opening of two new stores in the south east of Australia. One is in Springvale in Victoria, the other in Tempe in southern Sydney. Each new store creates unique opportunities to introduce sustainability initiatives. Some will be initiatives similar to those already introduced in other store locations around the world and others may be new Australian initiatives enabled by the latest technological breakthroughs. Each site presents unique conditions that will govern sustainability strategy and operating conditions for that site. Local governing authorities, regulations and planning conditions will also influence the approach to be taken at each location.

Possible inclusions for new sites are:

- solar hot water heating

- insulated membrane roofing material

- sun powered electricity (photovoltaics)

- LED and energy efficient lighting systems

- recycling and waste management

- community involvement programs

- comprehensive landscaping

- customer education on sustainable living opportunities.

# Case observations

The IKEA case outlines how a global organisation has embraced sustainability as an important core for its ongoing growth strategy. The 2009 IKEA Sustainability Report highlights key elements of how sustainability will be incorporated into policy and strategy. An extract from the introductory letter published in the report by global CEO, Mikael Ohlsson, reflects this commitment:

> We have a strong foundation to build on - it is in our culture to twist and turn established truths to find a new angle, a new idea, and to have the courage to try to do things differently. IKEA is obsessed with making more from less and we hate waste of any kind. This will continue to be our compass in years to come, and we will stimulate new thinking and innovation in our sustainability work.
>
> Innovation is needed to build sustainable solutions into the IKEA range and to tackle some of the global challenges society faces. We need to identify and use more resource efficient material and develop better solutions for reusing and recycling IKEA products once our customers no longer want them.
>
> I believe IKEA together with our customers, co-workers, suppliers and the rest of society can make a big difference.

IKEA began its approach to sustainability because its executives thought it was the right thing to do and also to cut costs by reducing waste. In recent times, there is growing realisation that the group can gain further business advantages by pursuing sustainability as an integral part of its core growth strategies.

It is useful to place IKEA's progress on achieving sustainability on a corporate sustainability model which identifies the phases that an organisation may pass through on the path to becoming a sustaining organisation (Dunphy *et al.* 2007). This model identifies the characteristics of each phase in terms of an organisation's approach to both human and ecological sustainability.

Phase one is *Rejection* when an organisation actively rejects the case for taking sustainable actions in either area – human or ecological. In phase two, *Non-responsiveness*, an organisation does not consider human or ecological sustainability actions as part of its own charter or domain of responsibility, regarding such things as irrelevant to business. Phase three is termed *Compliance* where an organisation pursues what could be seen as the current minimal sustainability requirements, such as conforming to occupational health and safety regulations and reducing emissions. An organisation is likely to follow this approach as part of its risk management strategy to avoid liability under laws that set down conditions for both human and ecological compliance. *Efficiency* is the term given to phase four of this model. Here there is a systematic approach to progressively reducing costs by eliminating waste in all areas of the operation including human resource management, procurement and marketing. In this phase, environmental issues are likely to be ignored if they do not generate recognisable cost savings and efficiencies. Phase five is known as *Strategic Proactivity* where both human and environmental strategies are considered vital to an organisation's long-term growth and successful survival. Here the organisation considers these elements as core business strategies that can lead to sustainable competitive advantage. Upon reaching phase six, the *Sustaining Corporation* goes further, accepting ongoing responsibility for the progressive upgrading of human knowledge and capital. It also becomes an active promoter of ecological sustainability values and seeks to influence industry members and the wider society.

After completing our recent interviews and reviewing relevant documentation for the purpose of writing this case study, it is our opinion that IKEA is currently beginning the move from a phase four corporation (*Efficiency*) to a phase five organisation (*Strategic Proactivity*). In the introduction to the 2010 IKEA Sustainability report, IKEA Group CEO, Mikael Ohlsson expresses a top level commitment to include sustainability in the future growth strategies of IKEA's international operations:

> IKEA has an important role to play in terms of taking responsibility for people and the environment. This is why sustainability is one of our four cornerstones in the new Group strategy: Growing IKEA together.

Our case research shows that IKEA has introduced numerous and varied measures aimed at improving the sustainability of its global operations. Since the early 1990s IKEA has announced policies that have been intended to reduce any negative environmental impacts of its operations and product offers. This has been both a direct initiative and also one conducted in partnership with its many suppliers. Through its e-learning policy, the group has also committed to educating key stakeholders on such issues as human rights, labour and working conditions, environmental management and corruption.

The launch of the IKEA social and environmental strategy in 2006 had a particular focus on the role and importance of leadership, competence, building and infrastructure, community involvement, energy and transport. This strategy has progressively resulted in ongoing initiatives in its existing global operations. It also forms the basis for setting core sustainability requirements and strategies for the new stores that are being planned and progressively opened in Australia and elsewhere. New store policy embraces key areas of waste reduction, recycling, energy conservation, social engagement, stakeholder initiatives, education and learning. As various technologies advance and become more cost efficient, sustainability standards also advance to form a baseline model for the next IKEA store.

As IKEA continues to adopt a more strategic orientation in this area, we anticipate a more coordinated approach to its sustainability initiatives. Sustainability is becoming an integral part of its core growth strategies rather than resulting mainly from grass roots localised initiatives as in the past. The text discussed above (Dunphy *et al.* 2007, pp. 177-9) provides guidelines that IKEA senior managers could follow in their quest to move effectively from an efficiency focus to strategic sustainability. These steps are:

*Step 1:* Top team elaboration of corporate goals that embrace the key elements of sustainable thinking and actions.

*Step 2:* The development and systematic alignment of measurement systems that show ongoing progress of how corporate actions are achieving sustainability objectives.

*Step 3:* Ongoing diagnosis of opportunities for gains in both sustainability and corporate growth.

*Step 4:* Implementation and diffusion of successful practices both internally and externally.

*Step 5:* Review, monitoring and alignment to ensure that sustainability initiatives maintain an alignment and coordination with the major strategic direction of the organisation.

We can see that initiatives are being taken in all these areas. It is also evident that IKEA has already some of the characteristics of organisations in the final phase of the model for it is already actively promoting both human and ecological sustainability in the wider community.

Josja Van Der Maas emphasised that IKEA's current global extended sustainability initiatives are not widely known by the public but represent a potentially powerful business advantage. With the increasing public awareness of the environmental crisis, more people are seeking to support companies committed to creating healthy social and ecological outcomes. IKEA intends to continue to be a leader in modelling how corporations can move toward sustainability and also in influencing other companies and community groups to make similar commitments.

# Appendix 5.1 – List of IKEA interviewees

## Name and position

Richard Ansell, *Property and Expansion Manager*, IKEA Australia

Stafano Brown, *Sustainability Manager*, IKEA Group (Retail)

Annie Chandler, *Social & Environmental Manager*, IKEA Australia

Cass Hall, *Former Marketing Manager*, IKEA Australia

David Hood, *Country Manager*, IKEA Australia

Jude Leon, *PR Manager*, IKEA Australia

Josja Van Der Mass, *Deputy Country Manager*, IKEA Australia. To be appointed as Store Manager to the new IKEA Tempe store.

# References

Dunphy D C, Griffiths A, Benn S 2007, *Organisational Change for Corporate Sustainability*, 2nd edn, Routledge, London.

IKEA website, <www.ikea.com.au>, viewed 16 November 2010.

UTZ website, <www.utzcertified.org/index.php?pageID=107>, viewed 15 November 2010.

# Case 6

# Indian clothing industry: Ethical and social responsibility dilemmas

STEPHEN CHEN

## Introduction

The Indian clothing industry is the second largest producer of textiles and garments in the world, with exports during 2007-2008 totalling US$ 9.68 billion. It supplies the Americas, the EU, much of Asia and the Middle East. It contributes to 8% of India's total exports, employs 6 million people, and is the second largest provider of employment in the country (Indiamart website). However, in recent years, the industry has been the subject of a number of investigations regarding its use of child labour.

In August 2010, *The Observer* newspaper in the UK reported that some of the biggest clothing retailers in the UK were at the centre of another major scandal involving sweatshops in India. *The Observer* investigation found workers at the Indian suppliers of Gap, Next and Marks & Spencer (M&S) working up to 16 hours a day, which is in flagrant breach of the industry's ethical trading initiative (ETI) and Indian labour law. A worker at one factory described how he would leave for work at 8am, start work at 9am and regularly work through to 10pm with two half-hour breaks – though sometimes they would go through to 2am the next day and be expected to return again later in the morning. Some workers claimed that they had to work seven days a week, that they were paid at half the legal overtime rate and that those who refused to work the extra hours had been told to find new jobs, a practice defined under international law as forced labour and outlawed around the world.

All three clothing retailers highlighted by *The Observer* have stated that they are totally committed to ethical trading and will not tolerate abuses in their supply chain. Gap has a code of conduct for its suppliers while Next and M&S are both members of the Ethical Trading Initiative (Appendix 6.1). Next said it had found the situation to be 'deplorable' and the chairman of the Indian company it uses has apologised and promised to make amends, blaming the increased demand for workers at the Commonwealth Games in Delhi, which left many factories short of staff. Gap said all its factories had to stick to comprehensive and strict standards, which it said were non-negotiable. It admitted that its staff had uncovered violations concerning excessive overtime and overtime wage payments in June, and that the supplier had been ordered to pay back all the outstanding money and reduce working hours to the legal limit. But it said that firing its supplier would only hurt the workers. Marks & Spencer admitted its supplier had been operating excessive overtime, but said it had acted quickly to tackle the problem. In 2009, M&S launched a five-year ethical trading plan, under the slogan 'Doing the Right Thing', and a spokesman for M&S said that it was essential that its suppliers upheld strong ethical standards and that this was a condition of doing business with the company. Despite these initiatives, child labour remains a problem in the Indian clothing industry. Two factors can be identified: working conditions in India and the nature of the clothing industry.

## The problem of child labour

According to UNICEF, one in six children aged 5–14 years old, or about 16% of all children in this age group, are involved in child labour in developing countries (UNICEF 2010). It is a particular problem in India, which has a large percentage of young children and the largest number of child labourers in the world (Figure 6.1). The share of workers aged 5–14 years in the total work force of the country is estimated to be 3.15% (International Labour Organization (ILO) website).

**Figure 6.1** Workers aged 5–14 years, India

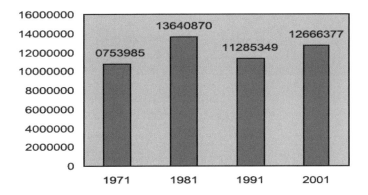

## Indian clothing industry

Child labour is most common in industries which rely on low-skilled workers, such as garments, footwear, carpet weaving, gem polishing, glass blowing, match works, brassware, electro-plating, lead mining, stone quarrying, and lock making (Kala 2006). The clothing industry is particularly well known for its reliance on child labour. In fact, the term 'sweatshop' has its origins between 1830 and 1850 as a specific type of clothing factory in which a certain type of middleman, the *sweater*, directed others in *garment making* (the process of producing clothing), under arduous conditions. Tailors or other clothing retailers would subcontract tasks to the sweater, who in turn might subcontract to another sweater, who would ultimately engage workers at a piece rate for each article of clothing or seam produced. The role of the sweater as middleman and subcontractor (or sub-subcontractor) was considered key, because he served to keep workers isolated in small workshops. This isolation made workers unsure of their supply of work, and unable to organise against their true employer through collective bargaining. The middleman made his profit by finding the most desperate workers – including immigrants, women and children – who could be paid an absolute minimum. Such a situation appears to be repeating itself in India.

**Figure 6.2** Structure of the Indian clothing industry

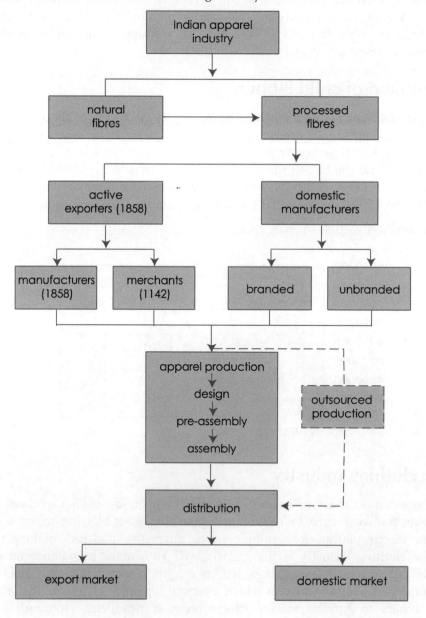

The clothing industry in India is divided into various sectors that make up a value chain which supplies to a majority of the world clothing market (Figure 6.2). The value chain starts with the natural fibre suppliers like cotton, jute, silk, etc. The natural fibres are supplied to fibre processers where the fibre is cleaned and spun into yarns. The processed fibre is then supplied to exporters and domestic manufacturers in the market. The domestic manufacturers consist of branded and unbranded apparel manufacturers; these producers only supply to the domestic market. The active exporters consist of apparel manufacturers that manufacture and export an item from under one roof. Merchants, on the other hand, receive the orders and outsource the production to smaller producers. It is in these smaller companies where most child labour has proven most difficult to eradicate.

The opportunities for such middlemen have increased as production in the clothing industry has become increasingly offshored and outsourced. Gary Gereffi, a leading sociologist working in globalisation in industries, has identified two types of global industry value chains: 'producer-driven' and 'buyer-driven' (Gereffi & Memedovic 2003). In producer-driven value chains, large, usually multinational, manufacturers play the central roles in coordinating production networks. This is typical of capital- and technology-intensive industries such as automobiles, aircraft, computers, semiconductors and heavy machinery. Buyer-driven value chains are those in which large retailers, marketers and branded manufacturers play the pivotal roles in setting up decentralised production networks in a variety of exporting countries, typically located in developing countries. Buyer-driven value chains are characterised by highly competitive and globally decentralised factory systems with low entry barriers. The companies that develop and sell brand-named products have considerable control over how, when and where manufacturing will take place, and how much profit is made at each stage. This pattern of trade-led industrialisation has become common in labour-intensive, consumer-goods industries such as garments, footwear, toys, handicrafts and consumer electronics.

The clothing industry value chain is organised around five main parts:

- raw material supply, including natural and synthetic fibres

- provision of components, such as the yarns and fabrics manufactured by textile companies

- production networks made up of garment factories, including their domestic and overseas subcontractors

- export channels established by trade intermediaries

- marketing networks at the retail level.

Major firms in the industry including retailers like Wal-Mart, Sears and JC Penney, athletic footwear companies like Nike and Reebok, and fashion-oriented apparel companies like Liz Claiborne, Gap and The Limited Inc., generally design and/or market—but do not make—the branded products they order. They are 'manufacturers without factories', with the physical production of goods separated from the design and marketing. Unlike producer-driven chains, where profits come from scale, volume and technological advances, in buyer-driven chains profits come from combinations of high-value research, design, sales, marketing and financial services that allow the retailers, designers and marketers to act as strategic brokers in linking overseas factories and traders, usually in less developed countries such as India, with main consumer markets in developed countries. Coupled with the relative ease of setting up clothing companies, this has led to an unparalleled growth in the number of garment exporters in the third world, such as in India.

## Production costs

Advantages for multinational clothing companies are clear. Wages in the garment industry in India range from 2.67 Rs to 48.14 Rs (US$1) per day (Table 6.1).

**Table 6.1** Textile industry wages in India

| STATISTICS: occupational wage surveys<br>Trend in average daily earnings by sex and age | | | | | | |
|---|---|---|---|---|---|---|
| MANUFACTURING SECTOR | Round no. | Average daily earnings (Rs.) | | | | |
| | | Men | Women | Adolescents | Children | Overall |
| Cotton textiles | I | 4.03 | 2.86 | - | 0.72 | 3.94 |
| | II | 6.06 | 4.76 | - | - | 5.69 |
| | III | 14.58 | 11.63 | 4.22 | - | 11.99 |
| | IV | 42.78 | 29.74 | - | - | 42.22 |
| | V | 78.12 | 73.24 | - | - | 77.77 |
| Woollen textiles | I | 3.59 | 1.55 | - | 0.87 | 3.47 |
| | II | 4.89 | 4.37 | - | 0.69 | 4.84 |
| | III | 13.76 | 8.22 | - | 0.82 | 10.04 |
| | IV | 35.83 | 35.96 | - | - | 35.84 |
| | V | 69.31 | 59.24 | - | - | 68.96 |
| Silk textiles | I | 4.09 | 1.68 | - | 0.78 | 3.89 |
| | II | 4.84 | 2.38 | - | 0.66 | 4.67 |
| | III | 10.42 | 7.10 | 2.84 | 1.33 | 9.61 |
| | IV | 30.63 | 25.15 | - | - | 28.93 |
| | V | 63.98 | 39.56 | - | - | 57.58 |
| Synthetic textiles*** | IV | 40.63 | 42.38 | - | - | 40.67 |
| | V | 62.38 | 40.86 | - | - | 61.72 |

| MANUFACTURING SECTOR | Round no. | Average daily earnings (Rs.) | | | | |
|---|---|---|---|---|---|---|
| | | Men | Women | Adolescents | Children | Overall |
| Jute textiles | I | 3.29 | 2.84 | - | - | 3.27 |
| | II | 3.79 | 3.61 | - | - | 3.78 |
| | III | 14.66 | 15.59 | - | - | 14.69 |
| | IV | 42.22 | 42.00 | - | - | 42.21 |
| | V | 89.73 | 85.99 | - | - | 89.63 |
| Textile garments | I | 3.23 | 1.69 | - | 1.20 | 2.67 |
| | II | 4.60 | 1.60 | - | - | 4.25 |
| | III | 11.27 | 7.12 | 2.38 | - | 10.29 |
| | IV | 34.01 | 18.61 | - | - | 25.85 |
| | V | 60.60 | 37.83 | - | - | 48.14 |

However, wages in the clothing industry need to be compared with wages with other less-developed countries (Table 6.2) and the alternatives. On the other hand, critics point to the huge disparity between the cost of production of clothing such as jeans and the prices at which such items are sold in shops in developed countries, some 3–5 times the cost of production (Table 6.2). The figures would suggest that major branded clothing manufacturers are capturing the bulk of the profits in the industry.

**Table 6.2** Average hourly apparel worker wages

| Hourly wage in US $ | |
|---|---|
| Bangladesh | 0.13 |
| China | 0.44 |
| Costa Rica | 2.38 |
| Dominican Republic | 1.62 |
| El Salvador | 1.38 |
| Haiti | 0.49 |
| Honduras | 1.31 |
| Indonesia | 0.34 |
| Nicaragua | 0.76 |
| Vietnam | 0.26 |

*Source*: Powell and Skarbek 2004.

Table 6.3 shows the basic production cost a company would pay to manufacture a basic pair of jeans that are retailed on an average for US$20–US$40 (Goldstein 2010).

Since India is a competitor, the average production cost of a pair of jeans is approximately the same.

**Table 6.3** Cost of production for a pair of jeans

| Cost component | Lesotho | Haiti | Nicaragua | China |
|---|---|---|---|---|
| | US$ | US$ | US$ | US$ |
| fabric | 3.07 | 3.27 | 3.20 | 2.80 |
| trims | 1.33 | 1.50 | 1.50 | 1.17 |
| wash/finish | 0.67 | 0.50 | 0.58 | 0.50 |
| labour | 2.00 | 1.67 | 2.00 | 1.67 |
| overhead/financing | 0.42 | 0.42 | 0.44 | 0.37 |
| duty | 0.00 (0%) | 0.00 (0%) | 0.00 (0%) | 1.09 (16.8%) |
| freight | 0.29 | 0.17 | 0.17 | 0.25 |
| total landed cost | 7.78 | 7.52 | 7.89 | 7.84 |

*Source*: Kurt Salmon Associates.

# Race to the bottom

Some economists have argued that such outsourcing to low-cost countries is a natural consequence of international trade. For example, in 1997, economist Jeffrey Sachs said, "My concern is not that there are too many sweatshops, but that there are too few" (Meyerson 1997). Sachs and other proponents of sweatshops cite the economic theory of comparative advantage, which states that international trade will, in the long run, make all parties better off. The theory holds that developing countries improve their condition by doing something that they do better than industrialised nations (in this case, they charge less but do the same work). Developed countries will also be better off because their workers can shift to jobs that they do better. These are jobs that some economists say usually entail a level of education and training that is exceptionally difficult to obtain in the developing world. It is also often pointed out that, unlike in the industrialised world, the sweatshops are not replacing high-paying jobs. Rather, sweatshops offer an improvement over subsistence farming and other back-breaking tasks, or even prostitution, trash picking or starvation by unemployment. An example is that after the *Child Labour Deterrence Act* was introduced in the US, an estimated 50,000 children were dismissed from their garment industry jobs in Asia, leaving many to resort to jobs such as 'stone-crushing, street hustling, and prostitution'. UNICEF's 1997 *State of the World's Children* study found these alternative jobs 'more hazardous and exploitative than garment production'. According to a November 2001 BBC article, in the previous two months, 100,000 sweatshop workers in Bangladesh had lost their sweatshop jobs. The sweatshop workers wanted their jobs back, and the Bangladeshi government was planning to lobby the US government to repeal its trade barriers to achieve this.

On at least three documented occasions during the 1990s, anti-sweatshop activists in rich countries have apparently caused increases in childhood prostitution in poor countries. In Bangladesh, there was a closure of several sweatshops that had been run by a German company, and as a result, thousands of Bangladeshi children who had been working in those sweatshops ended up working as prostitutes, turning to crime, or starving to death. In Pakistan, several sweatshops, including ones run by Nike, Reebok, and other corporations, were closed. This caused those Pakistani children to turn to prostitution. In Nepal, a carpet manufacturing company closed several sweatshops, resulting in thousands of Nepalese girls turning to prostitution (Balko 2004).

# Appendix 6.1 – Ethical trading initiative

## The ETI base code

**1.    Employment is freely chosen**

1.1    There is no forced, bonded or involuntary prison labour.

1.2    Workers are not required to lodge "deposits" or their identity papers with their employer and are free to leave their employer after reasonable notice.

**2.    Freedom of association and the right to collective bargaining are respected**

2.1    Workers, without distinction, have the right to join or form trade unions of their own choosing and to bargain collectively.

2.2    The employer adopts an open attitude towards the activities of trade unions and their organisational activities.

2.3    Workers representatives are not discriminated against and have access to carry out their representative functions in the workplace.

2.4    Where the right to freedom of association and collective bargaining is restricted under law, the employer facilitates, and does not hinder, the development of parallel means for independent and free association and bargaining.

**3.    Working conditions are safe and hygienic**

3.1    A safe and hygienic working environment shall be provided, bearing in mind the prevailing knowledge of the industry and of any specific hazards. Adequate steps shall be taken to prevent accidents and injury to health arising out of, associated with, or occurring in the course of work, by minimising, so far as is reasonably practicable, the causes of hazards inherent in the working environment.

3.2    Workers shall receive regular and recorded health and safety training, and such training shall be repeated for new or reassigned workers.

3.3    Access to clean toilet facilities and to potable water, and, if appropriate, sanitary facilities for food storage shall be provided.

3.4    Accommodation, where provided, shall be clean, safe, and meet the basic needs of the workers.

3.5    The company observing the code shall assign responsibility for health and safety to a senior management representative.

**4.    Child labour shall not be used**

4.1    There shall be no new recruitment of child labour.

4.2    Companies shall develop or participate in and contribute to policies and programmes which provide for the transition of any child found to be performing child labour to enable her or him to attend and remain in quality education until no longer a child; ["child" and "child labour" being defined below].

4.3    Children and young persons under 18 shall not be employed at night or in hazardous conditions.

4.4    These policies and procedures shall conform to the provisions of the relevant ILO standards.

**5.    Living wages are paid**

5.1    Wages and benefits paid for a standard working week meet, at a minimum, national legal standards or industry benchmark standards, whichever is higher. In any event wages should always be enough to meet basic needs and to provide some discretionary income.

5.2    All workers shall be provided with written and understandable Information about their employment conditions in respect to wages before they enter employment and about the particulars of their wages for the pay period concerned each time that they are paid.

5.3    Deductions from wages as a disciplinary measure shall not be permitted nor shall any deductions from wages not provided for by national law be permitted without the expressed permission of the worker concerned. All disciplinary measures should be recorded.

**6.    Working hours are not excessive**

6.1    Working hours comply with national laws and benchmark industry standards, whichever affords greater protection.

6.2    In any event, workers shall not on a regular basis be required to work in excess of 48 hours per week and shall be provided with at least one day off for every 7 day period on average. Overtime shall be voluntary, shall not exceed 12 hours per week, shall not be demanded on a regular basis and shall always be compensated at a premium rate.

**7.    No discrimination is practised**

7.1    There is no discrimination in hiring, compensation, access to training, promotion, termination or retirement based on race, caste, national origin, religion, age, disability, gender, marital status, sexual orientation, union membership or political affiliation.

**8.    Regular employment is provided**

8.1    To every extent possible work performed must be on the basis of recognised employment relationships established through national law and practice.

8.2    Obligations to employees under labour or social security laws and regulations arising from the regular employment relationship shall not be avoided through the use of labour-only contracting, sub-contracting, or home-working arrangements, or through apprenticeship schemes where there is no real intent to impart skills or provide regular employment, nor shall any such obligations be avoided through the excessive use of fixed-term contracts of employment.

**9.    No harsh or inhumane treatment is allowed**

9.1    Physical abuse or discipline, the threat of physical abuse, sexual or other harassment and verbal abuse or other forms of intimidation shall be prohibited.

The provisions of this code constitute minimum and not maximum standards, and this code should not be used to prevent companies from exceeding these standards. Companies applying this code are expected to comply with national and other applicable law and, where the provisions of law and this Base Code address the same subject, to apply that provision which affords the greater protection.

## Definitions

**Child:** Any person less than 15 years of age. If local minimum age law stipulates a higher age for work or mandatory schooling, the higher age applies. If local minimum age law is set at 14 years of age in accordance with developing country exceptions under ILO Convention No.138, this applies.

**Young person:** Any worker over the age of a child as defined above and under the age of 18.

**Child labour**: Any work by a child or young person younger than the age specified in the above definitions, which does not comply with the provisions of the relevant ILO standards, and any work that is likely to be hazardous or to interfere with the child's or young person's education, or to be harmful to the child's or young person's health or physical, mental, spiritual, moral or social development.

# References

Bheda, R, Naga, AS & Singla, ML 2003, Apparel manufacturing: a strategy for productivity improvement, *Journal of Fashion Marketing and Management*, 7(1), 12-22.

Gereffi, G & Memedovic, O 2003, 'The Global Apparel Value Chain: What Prospects for Upgrading by Developing Countries', United Nations Industrial Development Organization, Vienna.

Goldstein, J. 2010, 'Global Poverty And The Cost Of A Pair Of Jeans', <http://www.npr.org/ blogs/money/2010/03/how_much_does_a_pair_of_jeans.html>, viewed 18/6/10.

Indiamart, 'The insight view of Apparel Industry', <http://sourcing.indiamart.com/apparel />, viewed 18/6/10.

International Labour Organization, 'SRO-New Delhi', <http://www.ilo.org/public/ english/region/asro/newdelhi/download/publ/indus/census_rep.pdf>, viewed 18/6/10.

Jones RM 2006, *The Apparel Industry*, Blackwell Publishers, Oxford.

Kala T, 'Exploitation of Child Labourers in India', <http://www.wsws.org/articles/2006/ jun2006/indi-j08.shtml>, viewed 18/6/10.

Meyerson, A 1997, 'In Principle, A Case for More 'Sweatshops', *The New York Times*, 22 June 1997, http://query.nytimes.com/gst/fullpage.html? res=9B05E6D8103EF931A157 55C0A961958260, viewed 4 April 2008.

Powell, B & Skarbeck, D 2004, 'Sweatshops and Third World Living Standards: Are the Jobs Worth the Sweat?', Independent Institute Working Paper 53.

Radley B 2004, 'Third World Workers Need Western Jobs', Fox News, 6 May 2004, <http://www.foxnews.com/story/0,2933,119125,00.html>, viewed 18/6/10.

UNICEF (2010), 'State of the World's Children 2009', available at <http://www. childinfo.org>, viewed 18/6/10.

# Case 7

# Interface's approach to sustainability: Manufacturing green carpet

Wendy Stubbs

## Introduction

Ray C. Anderson founded Interface Inc. in 1973 in Atlanta, Georgia (US), to produce and market modular soft-surfaced floor coverings. The company is now the world's leading supplier of carpet tiles. Interface sells its products in over 110 countries, primarily in the business market, with manufacturing facilities in five countries. Interface is a publicly listed on the NASDAQ exchange and generates over US$ 1 billion of revenue annually.

In the mid-1990s, Ray Anderson refocused the company's strategy, aiming to re-engineer its industrial practices to include a focus on environmental sustainability without sacrificing its business goals. In 2000 Ray Anderson recognised that Interface's commitment to sustainability was incomplete and the critical missing factor was a genuine focus on people; social sustainability was equally important to implementing its strategy. A re-evaluation of sustainability took place through a collaborative discussion with employees. Sustainability was reinforced as the core value of the company and the social dimension of sustainability was put at the core of its vision alongside environmental sustainability and economic profitability.

Interface is widely acknowledged as a global leader in environmental sustainability. It was named by *Fortune* as one of the 'Most Admired Companies in America' and the '100 Best Companies to Work For', and has won numerous awards for its sustainability commitment and performance in Europe, America and Australia. Ethical Corporation recognised Ray Anderson with a Lifetime Achievement Award at the Responsible Business Awards 2010.

Interface has faced many challenges and barriers in its quest to become a sustainable company and, as a result, its business has undergone some radical changes in the past fifteen years. This case study describes three major, and inter-connected, components of Interface's strategy, which has resulted in significant business and sustainability outcomes:

1. **Integrating sustainability into the business**. Interface has been changing the way it does business by embedding sustainability into its core systems, processes and culture. As such, the commitment to sustainability has not wavered – even during financial downturns such as the global financial crisis (GFC) in 2007–2009.

2. **Collaborating with stakeholders**. Interface takes a holistic approach to implementing its sustainability vision and strategy. It takes responsibility for the environmental impacts of its full product life cycle (from sourcing raw materials, through manufacturing, distribution and disposal of product). Collaborating with its stakeholders is critical to achieving its sustainability outcomes.

3. **Measuring performance**. Doppelt and McDonough's (2010) research found that organisations making the most rapid progress toward sustainability have adopted effective internal measurement systems. Interface developed a number of measurement systems to track its environmental and social progress, to support more effective decision making in all areas of the organisation.

## Research methodology

Data were collected in the period 2003–2005 and in 2010 from in-depth interviews with staff engaged in sustainability initiatives, as well as from company briefings and other secondary sources. Ten 1–2 hour interviews were conducted in 2003–2004 and three interviews in 2010 with staff from operations, manufacturing, sales and marketing, finance and IT, services, sustainability management, executive management and the founder/chairman. Notes were taken from two presentations from Interface staff in April and August 2010. Secondary data were sourced from annual reports, quarterly earnings announcements, internal company documents, personal communications and the website. All interviews were recorded and transcribed and then coded to extract themes, using qualitative data analysis methods (Strauss & Corbin 1998).

## Background

On 31 August 1994, Ray Anderson delivered his vision of sustainability at Interface's global environmental meeting, in response to a request from his staff about what Interface was doing about the environment. His vision was formulated after reading Paul Hawken's (1993) book *The Ecology of Commerce* and him subsequently realising the level of environmental degradation caused by business and industry, particularly petrochemical dependent companies such as Interface. He saw himself as an environmental plunderer. He likened this epiphany to having a spear thrust into his chest, and sustainability became the driving force of his life and the mission of the company.

Interface's sustainability vision is:

To be the first company that, by its deeds, shows the entire industrial world what sustainability is in all its dimensions: people, process, product, place and profits — by 2020 — and in doing so we will become restorative through the power of influence.

To achieve this, Interface developed a model of the 'prototypical' company of the twenty-first century. This model embodies Interface's view of a sustainable enterprise: strongly service-oriented, resource-efficient, wasting nothing, solar-driven, cyclical rather than linear, and strongly connected to its constituencies (community, customers, and suppliers). The prototypical company seeks to go beyond complying with regulations, taking nothing from Earth's crust (lithosphere) that is not renewable and not harming Earth's biosphere (all living beings together with their environment) (Anderson 1998). To achieve this, Interface focuses its efforts on 'seven fronts' of sustainability, as summarised in Table 7.1.

The most challenging fronts are *closing the loop* and *renewable energy*. Closed-loop processing means 'getting off oil' – replacing the petrochemical derived raw materials with post-consumer recycled materials. It requires new processes, new technology. Suppliers and customers must be engaged to retrieve material for post-consumer recycling to close the loop. Renewable energy is a major challenge on a broad basis, as explained by one executive:

> *That links to how we run our factories, how we move ourselves around, how we move product around. I think that's because a large chunk of the things that we directly control ourselves, we've made lots of progress on. It's the areas that we need to tap into on a wider basis, wider initiatives, wider engineering breakthroughs, is where we're having most challenge. How to come up with some sensible ways to drive our factories, drive our products and ourselves... We continue to struggle to find viable [renewable] sources.*

In 2006, Interface launched Mission Zero – 'our promise to completely eliminate the negative impact our company may have on the environment by 2020' – to clarify and unify Interface's branding across all businesses worldwide and to succinctly express the company's mission to a global marketplace (Interface 2006).

**Table 7.1** Interface's seven fronts of sustainability

| Front | Description | Examples of products and programs |
|---|---|---|
| 1. Eliminate waste | Eliminate all forms of waste in every area of business. For Interface, this means redesigning products and processes to reduce and simplify the amount of resources it uses, so that material 'waste' will no longer be waste but instead will be remanufactured into new resources, providing technical 'nutrients' for the next cycle of production. | Entropy carpet: Random carpet design inspired by bio-mimicry (using nature's 'ordered chaos' as a design guide). QUEST (Quality Using Employee Suggestions and Teamwork): a program to reduce waste by 10% per year. |
| 2. Benign emissions | Eliminate toxic substances from products, vehicles and facilities. Interface has moved aggressively toward eliminating all its emissions into the ecosphere, striving to create factories with no smokestacks, effluent pipes or hazardous waste generated. | Cool Carpet™: World's first carbon neutral carpet. TacTiles™: a strong translucent sticker that binds carpet tiles seamlessly together which eliminates the issue of volatile organic compounds from traditional glue carpet adhesive. Reduces environmental footprint by 90%. |
| 3. Renewable energy | Operate facilities with renewable energy sources – solar, wind, landfill gas, biomass, geothermal, tidal and low impact/small scale hydroelectric or non-petroleum-based hydrogen. Interface seeks to ensure that by 2020, all fuels and electricity to operate its manufacturing, sales and office facilities will be from renewable sources. | Bentley Prince Street Solar Array: 128 kW photovoltaic array installed at the California manufacturing facility. Landfill Gas Project: convert naturally occurring methane gas from a landfill into a renewable fuel source for the manufacturing plant in Georgia. 8 out of 9 factories now use renewable energy sources. |
| 4. Closing the loop | Redesign processes and products to close the technical loop using recovered and bio-based materials. Interface is redesigning its processes and products to recycle synthetic materials, to convert waste into valuable raw materials, and to keep organic materials uncontaminated so they may be returned to their natural systems. | ReEntry®: post-consumer Nylon fibre is returned to Interface's fibre supplier where it, in combination with some virgin materials, is recycled into new Nylon for use in new carpet fibre. At the same time, the post-consumer vinyl carpet backing is recycled into new backing using Interface's Cool Blue™ backing technology. Bio-based material: making use of renewable resources such as bio-based fibres, to replace oil-based matt Zippered carpet: design carpet so that the easily detached from the fibre for recycling. |

| Front | Description | Examples of products and programs |
|---|---|---|
| 5. Resource–efficient transportation | Transport people and products efficiently to eliminate waste and emissions. Interface is working diligently to make its transportation more ecologically efficient and is participating in voluntary partnerships focused on reducing pollution and greenhouse gas emissions and offsetting the carbon dioxide emissions associated with its travel, including employee commutes. | Cool $CO_2$mmute™: Offsets employee commuting emissions. Cool Fuel™: Offsets business auto travel emissions. Trees for travel: Offsets business air travel emissions |
| 6. Sensitising stakeholders | Create a culture that uses sustainability principles to improve the lives and livelihoods of all of Interface's stakeholders – employees, partners, suppliers, customers, investors and communities. Interface believes that when stakeholders fully understand sustainability and the challenges that lie ahead, they will come together into a community of shared environmental and social goals. | Employee volunteer activities: Interface provides time off for employees to volunteer on local community projects, such as tree planting projects with local schools, and installing recycled carpets in homeless shelters or community centres. Interface Environmental Foundation: grants to conservation and environmental organisations and schools. |
| 7. Redesign commerce | Create a new business model that demonstrates and supports the value of sustainability-based commerce. Interface is creating new methods of delivering value to customers, changing its purchasing practices, and supporting initiatives to bring about market-based incentives for sustainable commerce, and has developed a model for other companies to chart their own evolution to a prototypical model of a 21st Century sustainable business. | Evergreen Lease: Carpet leasing service where a client pays a monthly lease fee and Interface takes back the product at the end of its life and recycles or repurposes the product. |

*Source:* Interface 2010 and interviews.

# Interface's approach to sustainability and change

## Integrating sustainability into strategy and business operations

Interface likens its quest to be sustainable to climbing a mountain higher than Everest, which it calls Mount Sustainability (depicted in Figure 7.1 – the placement of the fronts does not indicate that each front is addressed serially). Interface found that this journey requires continuous organisational learning and education, through partnering with thought leaders in sustainability who challenge Interface's thinking. As one executive stated, "don't do it alone". Interface's 'eco dream team' includes twelve sustainability visionaries, such as Janine Benyus (advocate of bio-mimicry), Paul Hawken, Amory and Hunter Lovins (co-authors with Paul Hawken of *Natural Capitalism: Creating the Next Industrial Revolution*) and Karl-Henrik Robert (founder of *The Natural Step*).

**Figure 7.1** Seven fronts of Mount Sustainability which Interface will conquer on its ascent

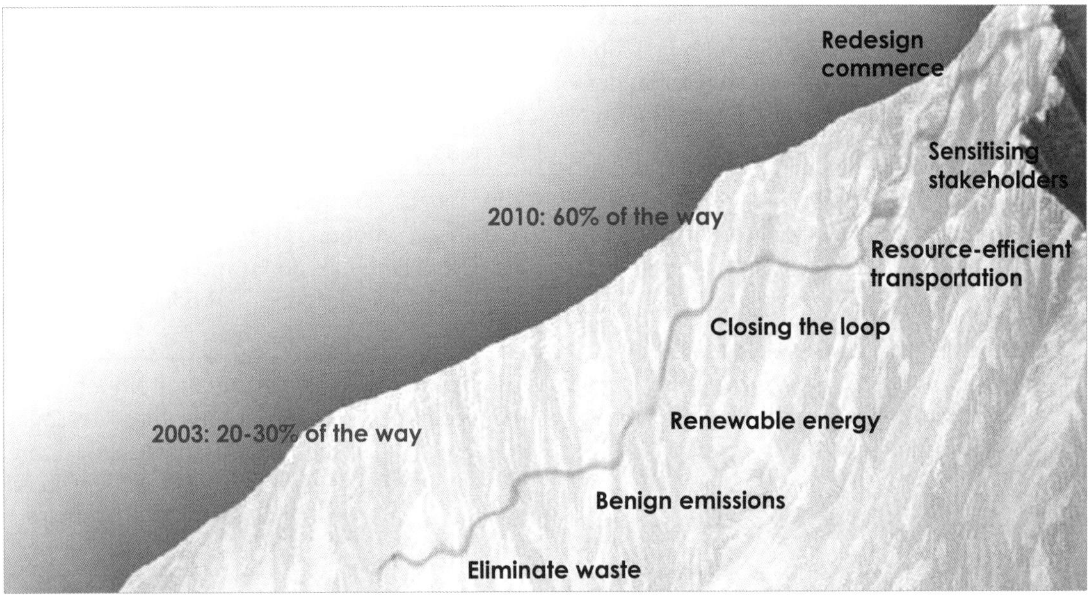

*Source*: Interviews and Interface website.

## A higher purpose to business

Interface began shaping its strategy using *The Natural Step*, which asks how nature would design an industrial system (see Table 7.2). Underlying the business strategy is the belief that there is a higher purpose to business than just making profits, as expressed by one executive:

> *The ultimate purpose of a corporation surely is something more than just making money. It has to make a profit to exist but it doesn't exist just to make a profit. Profit is an intermediate goal. It is the conventional way of keeping score. And there is nothing wrong with profit – we can't exist without it. But*

*what we are learning at Interface is that higher purpose for bringing people together.*

Another executive stated that when Interface initially started on the path to sustainability, the challenge was to reconcile two goals: achieving sustainability and maximising profits. Previously it only had one goal – to maximise profits. Now, one supports the other. People initially equated sustainability with higher costs and lower profit; having to spend time on something that was not profitable was distracting and was totally out of alignment with shareholder interests. The executive concluded:

*When you start to understand that it [sustainability] is not out of alignment with shareholder interests; when you know that, to use our term, it's a much better way to make a buck; you're there and you don't need convincing that it's right.*

**Table 7.2** Summary of The Natural Step

| System condition | Meaning |
| --- | --- |
| In a sustainable society, nature is not subject to systematically increasing concentrations of substances extracted from the Earth's crust. | This means substituting certain minerals that are scarce in nature with others that are more abundant and using all mined materials efficiently. |
| In a sustainable society, nature is not subject to systematically increasing concentrations of substances produced by society. | This means substituting certain persistent and unnatural compounds with ones that are normally abundant or break down more easily in nature, and using all substances produced by society efficiently. |
| In a sustainable society, nature is not subject to systematically increasing degradation by physical means. | This means drawing resources from only well-managed eco-systems, using those resources efficiently, substituting unnecessarily area-consuming activities with others and exercising general caution in all kinds of manipulation of nature. |
| In a sustainable society, human needs are met worldwide. | This means using all our resources efficiently, fairly and responsibly so that the needs of all our stakeholders – customers, staff, neighbours, people in other parts of the world, and people who are not yet born – stand the best chance of being met. |

*Source*: Adapted from Nattrass & Altomare 1999.

This approach also motivates and attracts staff. Many staff have worked in larger companies and manufacturing operations but are attracted to Interface because of the 'cultural feel'. One manager explained it as follows:

*For years and years I've been working on improving efficiency. And it's a much easier message to carry to the people on the shop floor that improving*

*efficiency and productivity reduces waste etc. and leads to improved [...]*
*BUT it also leads us to improve sustainability. Rather than being [...]*
*organisation which is purely about cost cutting and driving people harde[...]*
*harder to improve your profitability, here the people feel at least th[...]*
*contributing [to sustainability] and they're far more enthused about ma[...]*
*suggestions. Because they know they've got a purpose. The engagemen[...]*
*much easier to get on that message than purely on cost cutting.*

## Leadership

Leadership is critical. One executive stated that "it is not an intuitive way to run a business". Ray Anderson is a key driver of sustainability within Interface. Without his vision and leadership, Interface would not have undertaken, nor continued, its investment in sustainability. His burning passion for sustainability has kept the focus on sustainability even during economic downturns. Embedding sustainability 'in the culture' requires a commitment to the cause from management and all levels of staff, as well as having sustainability champions who can help educate staff and drive change. An important factor is clear and regular education and communication of Interface's sustainability vision to internal (employees) and external stakeholders (such as customers and suppliers). The founder summed it up in the following way:

> *I think you will hear that it [sustainability] is in the DNA of the company. It's who we are and what we are. It would take a whole set of changing of the guard to change that and then I think you would still have a huge push back from the people in the rank and file who know that this is a better way and come to work every day because they feel like the job is more important than just a job.*

Embedding sustainability in the culture is key to Interface's sustainability success. One executive explained that one of the best lessons they have learnt over the years is that "nothing else comes close to securing an engaged team of people – and that more than anything else has driven our success". However, Interface also learnt that "you can't dictate values – you need to leave room for people to come to it in their own way". A staff member who has been with InterfaceFLOR for about two years described a 'can do' attitude and an openness to new ideas and to challenging how things have been done in the past.

## Re-engineer thinking

Interface found that integrating sustainability into its business strategy and operations meant 're-engineering thinking'. Using sustainability as a design inspiration, Interface started designing products for recycling and dis-assembly/re-assembly and accepting natural defects rather than uniformity. For example, carpet traditionally is designed so that the backing does not easily detach from the carpet fibre. The engineers were asked to design a carpet that stays together while in use but at end-of-life easily comes apart for full recycling of the components.

Interface sent its designers out into the woods in Georgia and challenged then to integrate the principles of nature into design concepts for modular carpet – 'bio-mimicry'. The designers came back from their day in the forest with a sense that nature is organised chaos. Nature is a random and diverse system and each 'module' is slightly different in colour and shape. Out of this experience was born the Entropy™ line of carpet. Interface changed its manufacturing processes so that in one production run, the colour and design of each carpet tile would come out slightly different, which required the engineers to shift their thinking from a total quality management mindset (100% consistency and conformity). After launching the product, Interface found that, because it could lay the carpet tiles randomly, it not only saved on installation time and cost but it also reduced the amount of waste. In addition, it is easier and cheaper to make repairs because damaged or soiled tiles can be easily rotated – you don't have to worry about tiles being out of sequence in a pattern. Entropy is now Interface's most popular product.

## Start with waste

Interface's sustainability strategy can be summarised as: *take responsibility for the direct (own operations) and indirect (supply chain) impact; implement measures to reduce direct environmental impacts to the minimum possible, then offset the remaining direct and indirect impacts.* Interface estimates that over 90% of the $CO_2$ emissions associated with the life cycle of carpet occur outside its manufacturing processes, and it buys certified carbon offsets for its carbon neutral carpet.

In implementing a sustainability strategy, Interface found that it is best to start with waste. Eliminating all forms of waste from Interface's business operations has paid for investments in other sustainability initiatives. As one executive commented: "This is the banker, the one that pays for everything else we want to do". Waste is a cost, and eliminating waste directly affects the financial bottom line. To date, the cumulative saving from eliminating waste is US$ 480 million (1995–2010). Waste has a broad meaning to Interface, including waste from the production process as well as such things as customer complaints, rework, a wrongly-priced invoice or a bad debt. Interface staff initiated a number of waste reduction programs, but they are underpinned by Interface's strategy to move from a linear take-make-waste production model to a circular, closed-loop model where any waste is reused. This imitates nature, which does not generate any waste or toxicity as waste is used as food, or energy, for other parts of an ecosystem.

## The technology challenge

One of the major challenges for Interface in achieving its waste reduction program and Mission Zero is the ability to develop new technologies. As one manager commented:

> The biggest challenges are on the technology front. When we started this journey the technology didn't exist. We are a product of the first industrial revolution and we are trying to create the next industrial revolution with the cyclical processes, driven by solar energy and so forth. Linear being replaced with cyclical processes – take nothing, do no harm.

While most of Interface's large steps toward sustainability 'are tied with technology', Interface equally values the gains from small steps. On the manufacturing floor, "small changes like optimizing line speed and insulating

equipment to minimize heat loss have made a big cumulative impa
2010). All investments in technology must be financially justified vi
case. As one executive explained it:

> [Approval of] capital investment has to meet certain hurdle rates of return
> has to provide value back to shareholders. Investments in sustainability
> through the same process. They have to generate a return either through
> greater efficiencies of energy usage or material usage or add to customer
> engagement. But we're very mindful at the end of the day that we're a public
> company. We're not pursuing sustainability purely as a religion if you like,
> that must be done at all costs. It has to sit alongside the financial imperatives.

And Interface doesn't always get it right the first time, as illustrated by its
development of a bio-based fibre to replace nylon (and close the loop). Interface
developed a carpet using corn as a raw material. However, there were a number of
issues with using corn: many sources of corn were from genetically modified crops,
which concerned Interface; it was using a major food source, which could cause
problems for some developing countries; there were biodiversity impacts; and the
carpet itself was not as durable as nylon-based carpet. One manager remarked that
the corn story is "one of learning. It characterises our whole approach; often you
don't have all the answers but you're keen to try an idea and see where it goes".
Interface is now working with other bio-based forms of nylon that have superior
performance characteristics to nylon, and, "the supply-chain story is fantastic". One
source grows in desert areas and Interface sees the ability to create an agricultural
benefit for communities for whom agriculture has proved to be very difficult.
However, cost is the current challenge:

> We're a fair way down the track in terms of commercialisation. We've made
> carpet out of it [and] we're comfortable that it performs. Now it's about cost.
> Right now it's not economically viable to use it across the board to completely
> replace the virgin product. We can make sense of it technically, it's about how
> you commercialise it. Where do you put it, how do you use it, how to establish
> your supply chain, do you put it in everything, in 10%, do you put it in
> certain 'hero' products?

## Collaborating with stakeholders

A key learning for Interface is that it cannot achieve Mission Zero on its own; it
needs to collaborate with its stakeholders, particularly its supply chain. One
executive remarked that, "Ultimately, yes, the whole of the supply chain has to be
sustainable or we're not sustainable. We are our supply chain. Not many companies
think of it that way but we do". Interface takes responsibility for the impacts of its
products across the full life cycle and sets its boundaries widely – from "quarry/oil
well through to end-of-product-life". Interfaces uses a life cycle assessment (LCA)
tool to measure the environmental footprint of its products. LCA captures the
materials, energy and wastes involved in each phase of the product's life cycle: raw
material extraction and processing; internal manufacturing; transportation and
distribution; use, reuse, and maintenance; and, recycling or final disposal. Interface
needs to work closely with its suppliers to do this, as one manager explained:

*So you really need to collaborate with your suppliers and have a common purpose and the group has discounted some of its previous suppliers who were basically not of the same frame of mind; didn't have the same passion for sustainability. Some of our key suppliers are people we can really call partners. They've got the same objective; they see the future the same way as we do.*

For example, in the US, Interface has worked closely with its supplier Universal Fiber Systems to develop a process to take used carpet from the marketplace and develop a post-consumer recycled yarn, thereby closing the loop. This has "required investment on their part; required their Board to feel comfortable that there is a return [and] a growth opportunity for their business by partnering with us".

This approach has ramifications for supply chain partner relationships and strategy. It means that an organisation needs to review the sustainability of its suppliers' operations. This is a more holistic approach to achieving sustainability, which requires more cooperation between organisations in the supply chain to attack the core causes of environmental and social degradation at the source. It may mean the collaborative development of a sustainability strategy and plan for the whole supply chain – which is consistent with elements of Hart's (1995) 'product stewardship' (bringing stakeholders into the strategic process) and sustainable development (shared vision of the future) approaches.

Collaborating with stakeholders will become even more important as Interface moves closer to the peak of 'Mount Sustainability'. Interface anticipates that further reductions in its environmental footprint will likely come with collaboration from a growing number of stakeholders. President and CEO Daniel Hendrix stated that "The road ahead is more difficult and we'll not only need more innovative technologies, we also need collaboration from every direction – from suppliers, customers and the governments where we do business" (Interface 2010). This particularly applies to areas outside of Interface's core competencies:

*We are reasonably good in terms of materials, but that sits inside our core competency mix. When you look outside that, the development of sensible energy sources and how to apply that to your business, that is not within our core competency. So that dictates that you need to be working very closely with organisations and people that have those skills, have that knowledge and have the vision to want to get there.*

## Measure performance outcomes

While Interface does not produce a regular hard-copy sustainability report, measuring its sustainability impacts is critical to achieving its goals and strategy. The company did produce a report in 1997 and plans a Sustainability Report in late 2010 that will recap progress to date and map out the next ten years. Interface tracks its progress using a rigorous metric system established in 1996. These metrics enable the company to understand the impacts of its processes and products and to drive improvement. All the information is published on its website. Transparency and credibility is extremely important to Interface. By being transparent in its progress, it hopes to educate others and engage its stakeholders to join it on its sustainability journey. Interface has its results externally verified and certified and participates in the development of verification and certification standards whenever possible.

As well as standard financial indicators, which are required for a company listed on a stock exchange, Interface utilises four main performance measurement tools. QUEST (Quality Using Employee Suggestions and Teamwork) focuses on waste reduction. EcoMetrics measures environmental outcomes and SocioMetrics measures social sustainability initiatives (see Table 7.3 for 2009 results). EcoSense is a benchmarking tool for comparing the sustainability efforts of each of its facilities. It is like a competitive league table. It awards points for making progress on all the seven fronts. For example, points are awarded for riding a bike to work rather than driving a car and for sustainability projects with the local community.

**Table 7.3** Interface global sustainability performance as at December 2009

| Environmental metrics (EcoMetrics) | Result |
|---|---|
| Cumulative savings from global waste activities since 1995 ($million) | US$433 |
| Decrease in total energy consumption required to manufacture carpet since 1996 | 43% |
| Percentage of total energy consumption from renewable sources | 30% |
| Reduction in direct GHG emissions | 44% |
| Reduction in GHG emissions including offsets | 71% |
| Reduction in water intake per square metre of carpet (modular) since 1996 | 77% |
| Reduction in water intake per square metre of carpet (broadloom) since 1996 | 47% |
| Amount of material diverted from landfill since 1995 (million pounds) | 200 |
| Decrease in manufacturing waste sent to landfill since 1996 | 80% |
| Percentage of recycled or bio-based content in products worldwide | 36% |
| Safety—reduction in frequency of injuries since 1999 | 63% |
| **Social metrics for 2009 (SocioMetrics)** | **Result** |
| Number of employee/family social events worldwide<br>Lowest: 52 (2007) Highest: 243 (2003) | 69 |
| Average hours of training per employee<br>Lowest: 6.7 (2001) Highest:18.7 (2008) | 13 |
| Contributions to charitable organisations<br>Lowest: $252,900 (2002) Highest: $835,064 (2007) | US$331,000 |
| Employee volunteer hours in community activities<br>Lowest: 7,368 (2006) Highest: 18,775 (2008) | 9,057 |

*Source*: Interface 2010 and interviews.

In addition, Interface has developed a decision-making tool to help it compare the environmental impacts of different products and processes. The 'spider diagrams' map the environmental impacts of different options to help Interface design products and services with the lowest environmental footprint. For example, Figure 7.2 compares the impacts of using adhesive glue to lay carpet tiles versus using a new product called Tactiles (see Table 7.1). The small inner shading represents the 'footprint' of the Tactiles and the outer shading represents glue.

**Figure 7.2** Decision-making tool to assess environmental footprint

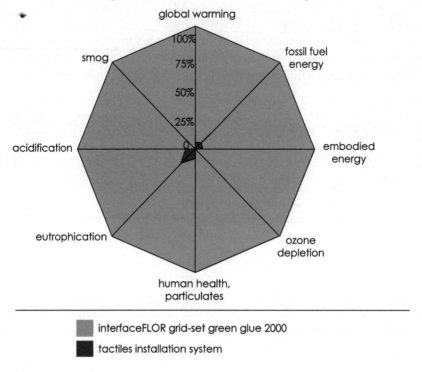

interfaceFLOR grid-set green glue 2000

tactiles installation system

*Source:* InterfaceFLOR 2009, p. 1.

# Results

Table 7.3 shows substantial progress on all of the environmental metrics from the mid-1990s to 2009. Progress on the social metrics has been patchier. This is partly influenced by economic and business conditions. The decline in 2009 from peaks in earlier years coincides with a global downturn in business due to the global financial crisis in 2007–2008. The other reason is that Interface has not focused as strongly on the social side as it has on the environmental initiatives. One staff member suggested this was an issue of prioritisation and time and it needs a more consistent approach. This was echoed by a senior executive:

> *This is the area around which we beat ourselves up regularly. We keep telling ourselves we are very strong in the environmental aspects and that the social side we have a less defined approach; a less coordinated approach. It's a much more organic process. You're talking about engagement with your own people; with the various stakeholders of the business on a local level... How do you put in place programs for social engagement across the wider community that has*

*continuity and consistency over months and years? We have come a long way but it's much harder to track the progress on a continued basis in this area.*

Overall, Interface believes it is about 60% of the way to its 2020 goal, compared to 20%–30% back in 2003 (see Figure 7.1). This has been achieved at the same time revenue and profits have grown. Globally, its net sales grew by 27% since 1996. In Australia, InterfaceFLOR is six times bigger than it was in 2002 and profits have increased ten-fold. The commitment to sustainability did not waver during the GFC because:

*… the business philosophy makes absolute business sense and that applies when times are tough and when times are good. We have never seen this as a cost to the business; we've seen it as a benefit to the business.*

Interface has found that it cannot charge a premium for 'green carpet', so the pricing remains competitive during boom times and downturns. Even though customers won't pay more for green carpet, Interface's strong commitment to sustainability gives them a competitive advantage, which helps them through a downturn:

*… we stayed quite true to the journey and I think we saw that also in the last downturn. We learnt that it was our commitment [to sustainability] that was a large part that sustained our customers' engagement throughout the 2000-01 downturn. We saw that come through in the post-recession research that was done earlier in this decade. There is no desire to water down our approach through the GFC.*

## Challenges and opportunities

Interface staff identified a number of challenges to achieving Mission Zero by 2020. The second half of the journey is harder, as the 'low hanging fruit' has already been picked. That is, large efficiency gains were made early on, but:

*… efficiency doesn't lead you to Zero. Ten per cent gain every year never comes down to Zero basically. You have to make a step change away from what you are doing to get to that Zero point. Definitely it is harder and harder and that's what we're doing at the moment. We're looking at a lot of options to be able to step change with regards to different materials, different technology, different processes.*

One of the biggest challenges is to 'get off oil'. For a company that is predominantly a petrochemical derived manufacturer, its challenge is to replace oil with renewable feedstocks and post-consumer recycled carpet. However, as one manager pointed out, this also presents a huge opportunity for Interface because the price of oil is predicted to increase as demand outstrips supply and as resources become scarcer:

*Sustainability itself is a huge opportunity … as the price of energy and materials go up and up as supply reduces, companies that have set themselves up to disconnect themselves from that source will be in a much better position. I think it's both a challenge to get there but once we're there it will be our biggest opportunity as well.*

Interface finds that the current property industry is difficult to work with. The industry has a 'demolish-and-dump mindset' and accessing large quantities of carpet tiles to recycle is difficult as the industry disposes of the material before Interface can become aware of it. Interface is trying to increase the visibility of its ReEntry® program (see Table 7.1) and redesign commerce (7th front), but:

> That's a different commercial model that says we might sell you the carpet but we want it back when you've finished with it. [We] start to see the built environment as the mine site for coming years for raw materials. Those two are linked. To improve the recyclability and recovery rates and the amount of feedstock we can access, we need to redesign some of the commercial practices in the property sector.

Competing with low-cost manufacturers, particularly in Asia, is challenging. Interface believes that being local is more sustainable as it can more effectively service its customers by being in touch with local needs, as one executive explained:

> We have a philosophy that we want to be making products as close to the customer as we can. Our supply philosophy is smaller scale factories closer to the customer, tapping into local communities, supporting local workforces, local supply and so on … [but] taking some of the [sustainability] initiatives that have been developed globally and making them work locally is a real significant challenge… The globalisation of business directly flies in the face of sustainability… If you look at our businesses around the world, our share of local markets is directly proportional to the proximity of the factory and of a source of supply. So the further away you go from the source of supply, the lower is the market share. It's as simple as that.

A major opportunity is to 'take the brand out to the consumer base' and sell carpet tiles in the residential sector. The challenges are: to grow market share while maintaining control of the Interface brand and the value it can offer to the consumer through a retail channel; and, overcome supply chain issues with a much larger sales volume and a different service model.

## Conclusion

The philosophy underpinning Interface's sustainability approach is that sustainability and profit-maximisation are mutually supporting goals, encapsulated by one executive: "You don't get traction if you don't believe and can't communicate that the profit motive and sustainability motive are mutually supportive. Frankly, if you get that, everything else falls into place". Interface's approach is to integrate economic, environmental and social goals into its business model rather than one of tradeoffs, where sustainability initiatives are cut (cost-savings) if profit targets are not met. However, all sustainability initiatives are subject to a business case, as per all capital investment decisions, so sustainability is not prioritised ahead of economic outcomes. On the flip-side, Interface can't charge their customers a premium for green carpet and generate higher profits from being sustainable. In this respect, stakeholders (shareholders and customers) are not asked to foot the bill for sustainability initiatives; instead these are funded from cost-savings through efficiencies gained from the sustainability initiatives (primarily waste reduction).

Interface's future sustainability efforts will increasingly focus around collaborating with stakeholders, since 90% of Interface's environmental impacts occur outside its operations, and acknowledging that 'we can't do it alone'. Reaching the peak of Mount Sustainability will require a whole-of-supply-chain approach, one that may require new organisational skills and capabilities.

# References

Anderson, RC 1998, *Mid-Course Correction*, Atlanta: The Perengrinzilla Press.

Doppelt, B & McDonough, W 2010, *Leading change toward sustainability: a change-management guide for business, government and civil society*, updated 2nd edn, Sheffield: Greenleaf.

Hart, SL 1995, 'A natural-resource-based view of the firm', *The Academy of Management Review*, 20, 986-1014.

Hawken, P 1993, *The ecology of commerce: a declaration of sustainability*, 1st edn, New York: HarperBusiness.

Hawken, P, Lovins, AB & Lovins, LH 1999, *Natural capitalism: the next industrial revolution*, London: Earthscan.

Interface 2006, Interface Launches "Mission Zero",<http://www.interfaceglobal.com/Newsroom/Press-Releases-(1).aspx>, viewed 18 August 2010.

Interface 2010, 'Energy Use Reduced Nearly One-Half; Sales up 27 Percent: Thirteen years in, Interface Inc. has Reduced its Footprint While Growing the Business', <www. interfaceglobal.com/Sustainability/Progress-to-Zero.aspx>, viewed 19 August 2010.

Interface website, <www.interfaceglobal.com/getdoc/224de860-bf76-4d2f-973c-80af60a 4addd /7-Fronts-of-Sustainability.aspx>, viewed 10 November 2010.

InterfaceFLOR 2009, Measuring Carpet's Environmental Footprint.

Nattrass, BF & Altomare, M 1999, *The natural step for business: wealth, ecology and the evolutionary corporation*, British Columbia: New Society.

Strauss, AL & Corbin, JM 1998, *Basics of qualitative research: techniques and procedures for developing grounded theory*, 2nd edn, Thousand Oaks: Sage Publications (first published 1990).

# Case 8

# Leighton Contractors: Becoming a sustainable organisation

TONY STAPLEDON AND ROBERT PEREY

## Introduction

The case study presented below was developed from research on organisational change for sustainability undertaken in Leighton Contractors. The research approach involved participant observation and interviews with executives and senior managers. The quotations in the case study are sourced from interview transcripts.

## The need for change

When Peter McMorrow was appointed as the Managing Director of Australian construction company Leighton Contractors Pty Ltd (Leighton) in 2004, he inherited a big company struggling to deliver on its potential. The company, a subsidiary of major Australian-based international Leighton Holdings Limited, had revenue of around AU$1.5bn, yet it had been unprofitable in the previous financial year due to a number of poorly performing projects, and was subject to ongoing contractual disputes and low morale.

However, when McMorrow retired from the position in 2010, the company had grown substantially – fourfold in employee numbers to some 10,000 people with annual revenues of AU$6bn and growing. Leighton was now highly profitable with nearly AU$9bn of work-in-hand up from AU$2bn in 2005, and its survey of employees showed high levels of employee engagement. The company's operations had also expanded. Its Construction Division was now supplemented by new divisions focussed on resources, telecommunications, industrial and energy, and investment and facility management. McMorrow and his team had set the company well on the path to success.

# Taking stock

In 2005 McMorrow suspected that Leighton's brand was in trouble so, soon after he took over, he commissioned research into the market's perception of Leighton. The research revealed widespread mixed views of Leighton ranging from an 'aggressive contractor' with 'brown-cardigan', or conservative, engineers through to a 'loyal and professional' company. He realised that poor market perceptions had to change to be more consistently favourable if the company was to be successful. McMorrow set out to redefine the Leighton brand; building a culture that both reflected the values and behaviours he wanted and increased alignment and direction for managers and employees.

Over a one-year period workshops were held with a cross-section of employees – in offices and on construction sites – to define a set of cultural values. The company's first Organisation Development (OD) Manager was appointed, and he worked with the leadership team of executives and 24 general managers to consolidate the feedback into five shared values:

- safety and health above all else

- respect for the community and the environment

- enduring business relationships

- our people are the foundation of our success

- achievement through teamwork.

The general managers were then tasked with socialising the values into the organisation. However, this proved difficult and the OD Manager decided that, to make the values more real for employees, it was necessary to be able to more fully articulate the behaviours that supported (and detracted) from each value. The initial work for this was done at a newly instigated leadership program where high potential employees workshopped what they considered those behaviours to be. Their suggestions were then brought together into three supporting behaviours for each value. For example, the behaviours for the value of 'safety and health above all else' are:

- look after our work mates as if they are family

- take action when we see a safety risk, and

- do not compromise safety for profit.

McMorrow promoted the values and behaviours by saying, "Our values define the way we work – it's not just what we do, but how we do it that matters". An extensive internal communications program was launched to inform employees about the values and behaviours and to ensure that they were carried into the company's day-to-day activities.

As well as being committed to creating a values-based organisation, McMorrow had strong personal beliefs about looking after the environment and the need to tackle climate change. Communication of these beliefs was symbolically important as management of the environment and greenhouse gas emissions posed challenging problems in Leighton's day-to-day work, which involved extensive diesel fuel consumption, disruption to land, and use of materials with high-embodied energy, such as concrete and steel. At the same time environmental concerns were becoming more important to clients, authorities and the public, as well as to existing and potential employees. During this period environmental issues were made more immediate through drought, water scarcity, and the high profile activities of Al Gore arguing the case for action to stop human induced large-scale damage to the natural environment. All these factors meant that sustainability was coming onto the agenda, and in 2006, the company decided to bring together data on its safety, health and environmental performance in order to publish its first sustainability report.

In 2007, McMorrow created a new position for a Group Sustainability Manager. Because sustainability was initially seen as predominantly an environmental issue and one that involved changing employee behaviours, the position reported to the Executive General Manager of Human Resources who was also responsible for the Environment function.

The new sustainability manager's first task was to understand and then explain how sustainability related to Leighton's business. His background was in business strategy and organisational economics, not the environment, and he set out to place sustainability within a broader strategic and operational framework. He began by developing a sustainability objective that resonated with Leighton; to be a sustainable organisation it was to be 'long lasting, consistently profitable and corporately responsible'. And, by demonstrating that corporate responsibility was consistent with the company's shared values, a path to sustainability was identified: 'By living our values – through what we do and how we do it – we can become a sustainable company and achieve our core purpose of transforming ideas to enhance people's everyday lives'. This idea of 'living our values through what we do and how we do it' brought together sustainability, the work that had been done on brand, values and behaviours, and the company's mission to enhance people's lives.

## Sustainability strategy

Following the critical first step of defining sustainability meaningfully for the company, a four-stream sustainability strategy was developed to bring about the change necessary for sustainability.

The first stream is aimed at establishing a culture in which the company's values (and so sustainability) become a consideration in everyday decision-making. This begins at recruitment and induction and carries through into corporate publications, leadership programs, and personal performance and development appraisals.

The second stream is to integrate sustainability (as distinct from identifying it as a separate function) into the business through consistent systems and procedures, changing the way people work and deliver projects and leading to best practice and continuous improvement.

The third stream uses sustainability to drive breakthrough thinking for innovation and business improvement, so creating opportunities for new business. One outcome of this stream has been a series of strategic alliances Leighton has formed with owners of renewable energy technologies.

The fourth stream builds engagement with clients, partners, suppliers, and those within the company to drive business improvement and resilience. For example, Leighton has a 30-year-long partnership with Caterpillar, and the two companies work together to improve heavy machinery design and the way it is used, so deriving mutual benefit.

**Figure 8.1** Four-stream sustainability strategy

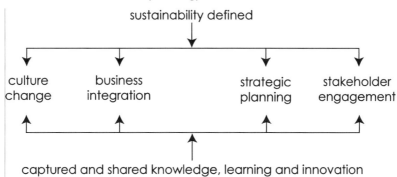

Knowledge and corporate learning underpin the overall strategy. Although the company does a lot of innovative things on its projects, its market and geographic diversity mean that learnings are often difficult to communicate beyond the project team. As a result, the company is working to strengthen the way knowledge is captured and shared so that innovation is off an ever-higher platform rather than a low base.

## Progress on sustainability

The take-up of sustainability has been highly dependent on individual managers' understandings of what is meant by sustainability in the context of the industry sector within which they work. While there is a consistent desire to change and 'do the right thing', there is still a need for more clarity about what sustainability means at the operational level. In part this is because each industry sector faces different sustainability priorities. For example, the buildings sector is very focussed on environmental (energy, water, materials use) efficiency, infrastructure construction often has significant environmental and community impacts, while mining has particular workforce safety and health, carbon emissions, indigenous employment and land disturbance issues.

Accordingly, Leighton found that sustainability implementation requires a flexible approach tailored to each industry sector. The consequence was that the sustainability team quickly adapted their implementation strategy to look for existing activities, processes and projects that they could link to the sustainability change agenda. The intention was to help people within the company bridge the gap between their existing knowledge and experience and good business practice consistent with sustainability principles.

For example in the infrastructure construction sector an explicit linkage was made to the principles and practices that underpin alliance contracts, which are becoming a preferred method of contracting in the sector. Alliance contracts are highly collaborative decision making structures where the key participants share the risks and returns of the project. They provide an easy entry to understanding sustainability because alliance performance and reward frameworks usually cover a range of project-critical economic, environmental and social issues.

What has also emerged during the implementation is an approach to sustainability - not unreasonable for an organisation flush with engineers - which takes a problem solving perspective to identifying solutions to environmental and social issues. Across all divisions there are consistently three main responses to sustainability challenges: compliance, efficiency and technologies.

Compliance plays an increasingly bigger role in operations as regulation becomes more stringent. The shift in sustainability thinking that has started in Leighton is, in part, a conscious move to go beyond strict regulatory compliance to seeking strategic opportunities that deliver concurrent beneficial socio-economic and environmental outcomes. For example, in one Western Australian project the Leighton team created new wetlands as an integrated component of a new highway. These extensions and enhancements to compliance are recognised as being important for both competitive and reputational advantage and are often referred to within the organisation as examples of how Leighton demonstrates 'its licence to operate'.

Another of these strategic responses is improved efficiency. One of Leighton's biggest operating costs, and also the area in which they can make the most significant carbon emissions reduction, is in diesel fuel consumption. The changes that Leighton have already implemented in their resources division include changing driver behaviour so that smoother driving techniques use less fuel, ensuring that road surfaces are better designed and maintained to reduce rolling resistance and redesigning equipment to make it lighter without compromising operational specifications. An example of the latter is working with equipment providers such as Caterpillar to provide lighter and stronger buckets for the mining equipment, which has led to a reduction in fuel consumption. However, available reductions using existing technologies are small compared to the targets that Leighton would like to achieve, and this leads to the third consistent response to sustainability issues – the search for new technologies.

By exploring and then supporting new technologies Leighton aims to open up new areas for business development and expansion as well as meeting changing client and society expectations for positive environmental and social action. An example of new technology that Leighton is investigating is in the field of low emissions energy where Leighton has forged partnerships with technology owners in gas, wave, solar thermal and wind energy, and in biofuels.

# Barriers

The changes to behaviours and practices that Leighton is endeavouring to make to influence sustainability performance take time. However, as with all major change programs, there are limits that an organisation confronts before it strikes barriers to adoption of sustainability. These barriers may be both internal and external.

Internally, there remains a need to better explain what sustainability means at an operational level; as one project manager explained: "You say sustainability, and people go all rubbery". There is a persistent view in some quarters that 'sustainability' is 'environment', yet, tempered with awareness that it is more than that – the triple bottom line. There is now work being done to equate sustainability with value for money, and this has struck a chord with project teams, particularly in the infrastructure sector.

Externally, organisations can only go so far before they come up against an industry/market problem that is articulated by 'if we do this we will no longer be competitive'.

This raises the question of how an organisation can maintain momentum internally for sustainability adoption without getting too far out of alignment with its markets. One aspect of Leighton's strategy for dealing with this balancing act is to play an active role in changing industry/market sustainability interpretations, processes and standards. For example, Leighton is involved with various industry bodies like the Green Building Council of Australia, the Australian Green Infrastructure Council and The Minerals Council of Australia, all of which have sustainability frameworks in place. Through this involvement, Leighton can contribute to shaping how industry approaches and self-regulates around sustainability. This new direction is another example of the flexibility of the corporate sustainability strategy, and its opportunistic evolution whilst retaining a clear intention to implement sustainability into the day-to-day practices of the organisation.

However, sustainability still faces hurdles. Many of the senior managers are deferring to the younger generation of engineers, as they are more in touch with sustainability and have more incentive to drive the necessary changes within Leighton. Yet the sustainability project is still relatively new and attempting such a large behavioural shift demands time and commitment from senior management. The importance of senior management commitment is the authority, certainty and predictability it provides for organisational change. The issue at the moment is that there is a change in the senior management team and people are not sure what new priorities may arise.

# A fork in the road

Peter McMorrow's retirement as Managing Director has brought changes to Leighton Contractors. A new Managing Director is in place, intent on building on his predecessor's legacy of a focussed divisional structure that pursues diverse markets while working within the strategic, policy and governance frameworks set at group level. There will be responsibility at all levels for delivering on sustainability and contributing to the Leighton core purpose of transforming ideas to enhance people's everyday lives.

A new sustainability strategy is being worked on. This aims to capitalise on the groundwork that has been undertaken over the previous period by strengthening sustainability governance, identifying priority strategic issues, and developing plans with defined responsibilities and measurable targets to address them. The second phase of the Leighton Contractors sustainability journey has begun.

# Case 9

# Ports of Auckland´s response to climate change related challenges

KATRIN HERDERING AND KATE KEARINS

## Introduction

In 2009, Ports of Auckland, New Zealand's leading port company, provided a vital point of connection with global markets. Its impact on the economy, society and the environment was predicted to strengthen because of its future growth aspirations. Client demand seemed to be shifting towards technologies that were more environmentally responsible, but there were still concerns about costs. If the company's strategies were not in tune with stakeholder expectations, its reputation could suffer and port business could go elsewhere in the country. Ports of Auckland's managers needed to take appropriate action on emissions reduction and carbon strategy in line with the company's vision to become "the best port in Australasia with world-class performance" (Ports of Auckland Limited 2008, p. 2).

This case was informed by background research on the company, other related businesses, and climate change and carbon strategy. A key source was an interview with Jim Harknett, Ports of Auckland's Chief Risk Officer in 2008 and material and ongoing correspondence from him through to the end of 2010, and duly attributed to him.

## Company background

Ports of Auckland is based in New Zealand's largest city, operating an international seaport and two domestic ports on the Auckland isthmus. In 2009, the 24/7 company employed 550 full-time equivalent staff to handle five commercial vessels on average, every day. Freight handling was the main business, with the company servicing "61% of the total upper North Island container trade by value, and handling cargo equivalent to 15% of the country's GDP – twice as much as any other New Zealand port" (Ports of Auckland Limited 2009, p. 2). Ports of Auckland also provided logistics for around 70 cruise ships that berthed at the city each year. There was some decline in overall profitability in 2009 due to the global recession and a significant decrease in vehicle imports (Record container traffic 2009).

Ports of Auckland's business was set to expand in order to meet the future growth requirements of the region and the nation. However, its managers aspired to increase business within the company's present footprint and without incurring unnecessary additional cost. The aim was to contribute to a high-speed, environmentally and socially sustainable logistics chain while balancing the requirements of all stakeholders. Managing Director, Jens Madsen's stated goal was to "extend the influence of Ports of Auckland's commitment to environmental efficiency and improved social outcomes beyond [the Port's...] facility, to the wider industry and supply chain" (Ports of Auckland Limited 2008, p. 1).

The company released its first stand-alone Sustainability Review in 2008. As well as covering port development planning, community and employment relations, health, safety and bio-security, there was a focus on greenhouse gas (GHG) emissions and energy use.

In 2009, all the shares in Ports of Auckland were owned by a statutory corporation called Auckland Regional Holdings, with dividend flows going to the Auckland Regional Council. That situation would change on 1 November 2010 with the formation of the new Auckland (Super) City Council when Ports of Auckland would come under its investment arm and be considered a Council Controlled Organisation.

## Climate change related challenges

Ports of Auckland managers were aware that extreme weather events could have adverse impacts. Rising sea levels would probably concern Ports of Auckland's waterside operations only in the very long run. However, severe weather could impact parts of the supply chain where assets were exposed to climate, including agriculture, forestry, tourism and insurance, with flow-on effects.

In 2008, New Zealand committed to the introduction of an Emissions Trading Scheme (ETS). The fundamental aim was to establish a price on GHGs to motivate a reduction in emissions. The ETS legislation did not affect Ports of Auckland directly because the company was not a large emitter. However, according to Jim Harknett, Ports of Auckland's Chief Risk Officer, the company was being encouraged to reduce its GHG emissions as an arms-length regional authority operation and because of increased public awareness surrounding environmental impacts. Ports of Auckland was a highly visible operation. Its main business was on prime waterfront real-estate in the heart of the country's business capital.

New Zealand's ETS was being implemented gradually, implicating several Ports of Auckland suppliers. The Ministry for the Environment (2008) forecast the costs of GHG emissions being handed on within the economic system, predicting an increase of electricity prices by approximately 5% from 2010 – even though around 70% of the nation's electricity was from renewable sources, mainly hydro. Carbon-based fuel prices were likely to go up by around 7% per litre from 2011. Manufacturers in other parts of the world with emissions legislation would also face higher climate-related costs. Some imported machinery would become more expensive.

There was wide-ranging discussion that customers might change their purchasing behaviour due to rising awareness of climate change and related issues. New Zealand´s high quality agricultural and forestry exports competed in far-away markets, including some where buyers´ environmental consciousness was high. New Zealand was frequently cited in concerns about food miles as its complete supply chain, from production through distribution to consumers on the other side of the world, appeared long; however, some researchers pointed out comparative anomalies (Saunders, Barber & Taylor 2006). Tourism to a place like New Zealand 'at the edge of the world' was also at some risk.

On the one hand, Ports of Auckland considered its "contribution to any product's total carbon footprint as relatively small", according to Harknett. On the other, as a critical point in the logistics chain, Ports of Auckland assessed its own footprint and saw its signalling of efforts to improve its footprint in relation to the wider industry as important given its stated vision and growth aspirations. Ports of Auckland wanted to demonstrate it was part of a more energy-efficient supply chain. The Port of Tauranga some 150km south-east was considered to be a competitive threat.

# Ports of Auckland´s greenhouse gas emissions

## Measurement of emissions

Ports of Auckland was one of the first ports in Australasia to measure and audit its GHG releases and carbon footprint. It did so across its three operational sites, following the Greenhouse Gas Protocol, distinguishing between direct and indirect emissions of the six GHGs, and subdividing emissions into different scopes (The Greenhouse Gas Protocol Initiative 2004). Scope 1 emissions resulted mainly from diesel combustion in Ports of Auckland-owned equipment. Scope 2 comprised indirect emissions caused by the company's use of electricity, a percentage being considered as deriving from natural gas or coal-fired power stations. Scope 3 included all other indirect emissions resulting from the company´s operations, but occurring from sources not owned or controlled by the company. Ports of Auckland reported these within narrow boundaries and took into account, for example, kilometres travelled by containers between the Auckland seaport and its inland port on vehicles driven by contractors. It did so because container movement was "driven by the strategic desire to improve the logistics chain in the Auckland Region and the emissions are material when compared to Scope 1" (Harknett 2010, p. 4). It also took account of Scope 3 emissions from land reclamation activities undertaken by contractors, as it considered reclamation "a vital activity … in the development of the port in the long term" (Harknett 2010, p. 4). A summary of GHG emission sources for Ports of Auckland appears as Table 9.1.

**Table 9.1** Emissions inventory summary (figures rounded)

| Type of emission | Tonnes $CO_2$-equivalent | |
|---|---|---|
| | **2009** | **2008** |
| **Total direct (scope 1) emissions** | **10,150** | **12,000** |
| **Total indirect (scope 2) emissions** | **3,250** | **5,000** |
| **Indirect (scope 3) emissions:** | | |
| - Inland port kilometres converted to diesel | 1,710 | 930 |
| - Air travel, taxis, mileage allowance | 90 | 170 |
| - Embodied emissions from reclamation – one time | 7,000 | 3,000 |
| **Total indirect (scope 3) emissions** | **8,800** | **4,100** |
| **Total emissions** | **22,200** | **21,100** |

*Source*: Harknett 2010, p. 3.

The year 2008 was set as a baseline against which the results of subsequent actions to increase/decrease emissions could be assessed. The company underwent the scrutiny of external auditor PricewaterhouseCoopers to ensure an accurate measurement procedure. The 2009 figures above were still being finalised at the time of writing.

## Management of emissions

Work done ahead of the detailed carbon inventory enabled Ports of Auckland to concentrate on actions that resulted in the most cost-effective GHG reduction. The biggest percentage of emissions was directly linked to diesel and electricity consumption. Three principal energy users were the company's marine floating plant (tug and pilot boats), cargo-handling machines (cranes and straddle carriers at the container terminal) and refrigerated containers.

From 2006 to 2008, absolute electricity consumption increased by 14%, but relative usage per 20-foot equivalent container unit (TEU) declined by 7%. In 2008, absolute diesel usage decreased for the first time in many years, even with considerable volume expansion. Between fiscal year (FY) 2006 and FY 2008, relative usage went down by 21% (see Table 9.2).

**Table 9.2** Energy use indicators 2006–2008

| Energy use indicators | FY06 | FY07 | FY08 | % change 2006 - 2008 |
|---|---|---|---|---|
| Electricity used by port operations (kWh) | 23,916,123 | 26,255,982 | 27,368,446 | 14% |
| Electricity used by port operations (kWh) per TEU | 34.86 | 33.96 | 32.54 | -7% |
| Diesel used by port operations (litres) | 3,519,950 | 3,608,034 | 3,413,842 | -3% |
| Diesel used by port operations (litres) per TEU | 5.13 | 4.67 | 4.06 | -21% |

*Source*: Ports of Auckland Limited (2008), p. 10.

Reduction in fuel consumption was realised in part through substantial capital expenditure on cargo-handling machines. In 2008, the company invested in 10 additional hybrid diesel-electric straddle carriers for stacking and transporting containers. The machines produced 90% fewer particle emissions, utilised 20% less diesel, and made less noise. Within a broader energy-saving plan, the company renewed workshop illumination and compressed air systems, realising a 25% energy reduction.

Lowering electricity and fuel costs would become more significant as energy prices rose. Reduced energy consumption automatically decreased the company´s GHG emissions, which in turn made the supply chain more efficient and contributed to minimising the carbon footprint of cargo handled by the company. Going forward, environmentally-conscious customers, such as importers or exporters, might consider this behaviour in their purchasing decisions, and the company stood to gain an advantage over competitors who did not do similarly.

Harknett admitted that investment in eco-friendly technologies had been modest in comparison to some overseas ports, but predicted it to increase. Due to recent tough economic times, the company had to deal with an expenditure restraint. It was important to identify the right investment options and to consider their impacts over time.

## Potential supply chain influence

### Dialogue with infrastructural providers

As Ports of Auckland contributed only a relatively small slice of the total GHG emissions of the entire supply chain, there was greater reduction potential in the carbon footprint of related industries. Cooperating with partners along the chain was touted as the way forward. Ports of Auckland had discussions with ONTRACK (rail network), KiwiRail (rail operator), the Auckland Regional Transport Authority and the Auckland Regional Council in an attempt to boost investment in rail infrastructure to one of its ports. The aim was to elevate the percentage of port cargo moved by rail from 12% to 30% thus building a more eco-sensitive and efficient supply chain, according to Harknett.

Although dialogue with infrastructural providers was a move in the right direction, arguably the company could do more by influencing its business partners upstream and downstream in the chain. The supplier-side comprised among others the producers of port equipment and electricity generators. Shipping lines, importers and exporters belonged to the customer-side (see Figure 9.1).

**Figure 9.1** Simplified supply chain at Ports of Auckland, including infrastructural providers

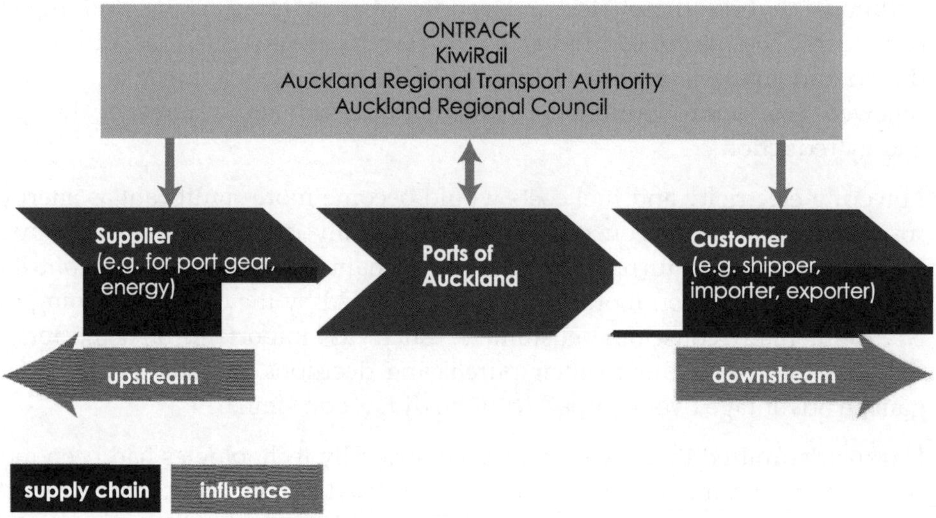

## Environmental monitoring and environmental collaboration

There were two further practices that Ports of Auckland could consider in relation to supply chain integration.

**Environmental monitoring** involved the use of arms-length, market mechanisms to shape other companies´ procedures through collecting and assessing their environmental disclosure by means of publicly-available sustainability reports, questionnaires or audits (Klassen & Vachon 2006).

On the supplier-side, the company could use its buying power to exert influence on contractors. Sustainable purchasing guidelines could be implemented. When bargaining with suppliers, the company could elect to apply key performance indicators. Furthermore, it could assess supplier performance on an environmental scorecard that included GHG emissions. A company like Ports of Auckland might find contractors with a significantly smaller carbon footprint. Efforts to decrease the wider industry´s GHG emissions by forcing eco-efficiency procedures up the supply chain sound easier in theory than in practice, however. A supplier for whom there is no alternative could refuse to improve its carbon footprint. Suppliers might struggle with a confusing range of stipulations from different purchasers.

Ports of Auckland could also attempt to influence the downstream side of the supply chain. It could seek to educate shipping line customers by enforcing standards concerning fuel emissions at its ports. The 'Environmental Ship Index', a program developed by the International Association of Ports and Harbours, measures the emissions of ocean-going vessels and honours ships that comply or fall below certain emissions standards (World Ports Climate Initiative 2009). Although the scheme is optional, willingness to comply may increase. A 2008 UN report revealed that "shipping emissions will become one of the largest single sources of manmade carbon dioxide" (Vidal 2008) – hence the sector will attract increasing attention. A port company with high aspirations would not want to be associated with the world's worst-polluting maritime vessels.

**Environmental collaboration** would see the company engage directly and devote its own resources to advance the environmental practices of partners through cooperative meetings concerning the environment and information-sharing relating to environmentally-sensitive product design or process improvement (Klassen & Vachon 2006).

Such collaboration could encompass partnering in the development of greener products and processes. Ports of Auckland could share its knowledge about successful tools applied for environmental enhancement, such as GHG measurement. It could also share know-how about effective and fuel-efficient use of equipment which could facilitate the progress of more environmentally-responsible innovation.

Another way to collaborate with customers could be to support the reduction of their emissions. Ports of Auckland could assist shipping lines by boosting the availability of on-shore power supply for vessels at the port. Most ships combusted fuel to provide full operational power when anchored; therefore, this measure could reduce GHG emissions (Miller 2008).

## The way forward

Ports of Auckland management made a start, but good decisions about carbon strategy are needed, including the identification of what should be its next focus. As with many sustainability-related decisions, trade-offs need to be fully considered along with cost-benefit considerations that focus on what would be good for the company, the industry and beyond – in the short and long terms.

## References

Harknett, J 2010, 'Draft emissions inventory report: 1 January 2009 - 31 December 2009', Ports of Auckland Limited.

Klassen, RD & Vachon, S 2006, 'Extending green practices across the supply chain: The impact of upstream and downstream integration', *International Journal of Operations & Production Management: Supply chain management theory and practice*, 26 (7), 795-821.

Miller, JW 2008, 'Ship shape: Ports spearhead drive to clean up shipping', For The *Wall Street Journal*, <http://www.wpccrotterdam.com/news/7>, viewed 31 March 2009.

Ministry for the Environment 2008, 'Factsheet 26 - Small and medium-sized businesses and the emissions trading scheme', <http://www.mfe.govt.nz/publications/climate/ emissions-factsheets/factsheet-26.html>, viewed 20 May 2009.

Ports of Auckland Limited 2008, 'Sustainability Review 08', <http://www.poal.co.nz/news_media/publications/POAL_sustainability_review_2008.pdf>, viewed 4 April 2009.

Ports of Auckland Limited 2009, 'Sustainability Review 09', <http://www.poal.co.nz/news_media/publications/POAL_sustainability_review_2009.pdf>, viewed 27 April 2010.

'Record container traffic, but vehicle slump hits Ports of Auckland profits' 2009, New Zealand Herald, March 10, <http://www.nzherald.co.nz/business/news/article.cfm?c_id= 3&objectid=10560912>, viewed 20 May 2009.

Saunders, C, Barber, A & Taylor, G 2006, 'Food miles – Comparative energy/emissions performance of New Zealand's agriculture industry', Lincoln University Agribusiness and Economics Research Unit Research Report No. 285.

The Greenhouse Gas Protocol Initiative 2004, 'A corporate accounting and reporting standard', <http://www.ghgprotocol.org/files/ghg-protocol-revised.pdf>, viewed 24 May 2009.

Vidal, J 2008, 'True scale of $CO_2$ emissions from shipping revealed', *The Guardian*, 13 February, <http://www.guardian.co.uk/environment/2008/feb/13/climatechange.pollution>, viewed 31 March 2009.

World Ports Climate Initiative 2009, Environmental Ship Index, <http://www.wpci.nl/projects/environmental_ship_index.php>, viewed 6 June 2009.

# Case 10

# State of Grace: Can death be sustainable?

Eva Collins, Kate Kearins and Helen Tregidga

## Introduction

New Zealanders Deborah Cairns and Fran Reilly founded State of Grace, an alternative family-directed funeral business in 2006. Inspired by a desire to create a paradigm shift in the funeral industry, the business offered socially and environmentally friendly alternatives for clients, including the involvement of family, keeping the deceased at home without chemical embalming and using natural products and eco-caskets. Ultimately, Deborah and Fran sought more natural alternatives to the current approaches to burial and cremation.

By August 2007, State of Grace had been operating in Auckland, New Zealand, for just one year and had performed 25 funerals. Deborah and Fran were trying to find the balance between growing the business, meeting their family commitments and staying true to their sustainability values. "Right now we are desperately trying to find premises. We're overflowing in my garage at home and into the hallway – there are caskets everywhere", Fran said.

## Research methods

This case study includes interview data from individual discussions and emails with Deborah Cairns and Fran Reilly in 2007 – and in 2010 for the epilogue[10]. It also includes data from government, industry association, business and other websites, and media articles, as cited. The case study reflects discussions we had with funeral industry participants, friends, colleagues and students – whose fascination with the topic and interest in aspects of the case further encouraged us.

---

[10] The assistance of Deborah Cairns and Fran Reilly is gratefully acknowledged.

## Getting started

Deborah's involvement in sustainable businesses went back to 1985/86 when she, her husband and another partner started an organics beverage company (Phoenix Organic). Deborah always had a sustainability vision. "I would not have wanted to be a part of a business that was not sustainable, ethical, doing some good. It's the only way that I'd work". As part of that conviction, State of Grace was a member of New Zealand's Sustainable Business Network (SBN), a network of businesses with an interest in sustainability. The SBN defined a sustainable business as, "the integration of economic growth, social equity and environmental management, both for now and for the future (Sustainable Business Network 2008). Deborah had some funding and some freedom to support her vision because in 2005, Phoenix Organic was sold to juice-maker Charlie's Group for NZ$10 million[11]. Fran's investment was from personal funds. She came on board, interested to see how the sustainable business model might work.

Deborah and Fran started by personally meeting everyone they would be doing business with in the industry. They took morning tea to the crematoriums. They met with funeral directors. "It has been hugely helpful in our day-to-day work", Deborah said, "not to mention helping to dispel the notion that we were two hippies". The personal touch worked with both suppliers and some competitors. For example, one funeral director allowed them to use his facilities. "We pay for his services, but it has been wonderful to have his advice and support where needed".

The two women took turns at being on call, which generally worked out fine. But now more people had come to know about State of Grace – and a single week in June 2007 saw the women managing five funerals and literally running between clients and their own families.

## A business based on temporary intimacy

Deborah and Fran had started out with a bold idea: operate a for-profit business based on creating intimate, but temporary, relationships with strangers often at one of the most private and stressful moments of those people's lives – and without any formal training themselves. Deborah articulated their vision:

> Our big vision was helping to create some sort of paradigm shift really, away from giving up this important part of our lives to other people. What we're doing is encouraging families to be as fully involved as they can with a person who's died. And that can be from very practical things like washing and dressing, right through to just accompanying someone, and not leaving them on their own in a strange place.

State of Grace offered all the services associated with our common rituals of death – but with an emphasis on socially and environmentally friendly alternatives wherever possible. The two women organised such things as newspaper notices,

[11] According to http://eh.net/hmit/exchangerates/, accessed 23 October 2007, $10 million NZ in 2005 was the equivalent of $7,148,010 US.

funeral celebrants and catering, as well as working with family or friends of the deceased, and doing what they referred to as "body care".

Body care was a large part of State of Grace's services and the one with the most physical intimacy. Embalming[12] was regarded as unnatural, so keeping the body, particularly the vital organs, cool was essential to delaying decay until after the funeral. Fran explained:

> We don't pack the orifices. That's always the first question people ask, and no, we don't. We wash down the body. Usually when they've been at home, they're already clean anyway, they're lovely and clean. And often they've had a long illness and probably haven't eaten anything for days, and they're fine. So we'll just wash them down, and we'll wash their hair. Because if somebody is really sick and in pain, that's just too much for them ....and so we'll wash their hair in the bed, and moisturise their face just to stop it drying out. And we dress them. And yes, ice packs underneath, which need changing regularly. Some families are fantastic and want to do that themselves, and they do a really good job. Some families really want to do it and do a really bad job. Some families just don't want to do it at all, so we do it. It's amazing how quickly those slicker pads melt....

Deborah admitted that she did not love all aspects of body care.

> I love the easy bodies that just need bathing and dressing, but struggle with the leaking or oozing, or the ones with strong odour. These would be better cared for by a hospice nurse. But generally, I really do love the act of bathing and caring for a person who has just died. I get a sense of who they were, I feel trusted, and I know that I do a loving and respectful job.

The analogy with midwifery was obvious to the two women. Fran explained,

> We get a similar thing that midwives get too, because you've come into somebody's life at a very personal, emotional level, they then think we're their new best friends. And it's hard for them to let go, and they're asking, 'when are we going to see you again?' And of course we can't do that.

Helping families let go of them when the services were over was one reason that Deborah and Fran ultimately wanted to develop a holistic centre that they could share with celebrants and grief counsellors.

Deborah and Fran thought that many of their clients chose State of Grace because it was run by women. Fran noted feedback they had received indicating many people would not want a man touching their mother, wife, or daughter. "Compassion and communication with grieving family members is essential. It's possible a woman may handle this a little differently to a man, who knows?" Deborah was even more certain: "We're mothers. We know about caring, and holding, and how to behave

---

[12] Embalming of the body involves disinfecting and preserving fluids being distributed through the body's arterial system by skilled personnel, and the posing of facial features. Preparation also includes washing, dressing, hairdressing and restoration of natural skin colour. (Funerals New Zealand 2007)

around sadness". "People are starting to reclaim their right of choice", Fran said. "We clearly offer them a service which can satisfy their own personal needs and allows them fairly unlimited involvement if they wish. I do think the fact that we are women has a safe and nurturing feel to it".

As well as this focus on the personal side, State of Grace was working to minimise the environmental impacts of traditional burials and cremations and for fair pricing for clients, as discussed later. They believed that a sustainable business model would eventually result in sufficient economic return. The main industry players in New Zealand operated on a more traditional basis, however.

## Daring to do things differently in a traditional industry

State of Grace was trying to break into a funeral industry that catered for most of the 28,460 deaths registered in New Zealand in the year to March 2007. Deaths were predicted to increase to about 29,490 in 2011, 33,840 in 2021 and 58,630 in 2051. Median age at death in March 2007 was 76.6 years for males and 82.3 years for females. Only 5.3% of deceased were under 40 years. The 7,157 deaths of residents in the Auckland region where State of Grace was located (and home to approximately one third of New Zealand's population of almost 4.2 million) accounted for only one quarter of the nation's deaths due to the region's relatively young age structure (Statistics New Zealand, 2007).

Burial had traditionally been the most common choice for families in New Zealand, with burials limited by law to official cemeteries or Maori burial grounds[13], apart from a few exceptional circumstances (FDANZ 2007a, p.1). The purchase of a plot, as well as an interment fee covering digging, maintenance and a memorial stone or plaque, meant that in most cases cremation was a cheaper option (Funerals New Zealand 2007a) with over 60% of families choosing it in 2007 (Funerals New Zealand 2007b). Cremation was no longer frowned upon by Christian religions (Funerals New Zealand, 2007b). It was not common in New Zealand to inter deceased persons in above-ground mausoleums.

Established in 1937, the Funeral Directors Association of New Zealand (FDANZ) was the largest national association of funeral directors in the country, with its members representing around 85% of funeral directors throughout New Zealand (FDANZ 2007b, p. 1). It claimed to conduct about 90% of all funerals (Griefcare 2007a) in New Zealand. Acceptance as a member of FDANZ signalled a commitment to adhering to a code of ethics and conduct, providing "a thoroughly professional and high quality service" and undergoing regular monitoring by the association (FDANZ 2007b, p. 1). Those on the FDANZ Register of Funeral Directors were required to have a national qualification in funeral directing, participate in on-going training and have (at least) a minimum standard of premises and facilities (FDANZ 2007b, pp. 1–2). Members paid a fee per funeral directed.

---

[13] Maori are New Zealand's indigenous people. They generally bring their dead to lie on their marae (traditional meeting places) and bury them in their urupa (tribal cemeteries).

State of Grace did not meet all of these membership requirements. Deborah explained, "We can never become members because you have to do the training, and to do the training, you have to work for an established funeral director for two years. I think that if there are enough of us [eco-focussed industry members], we need to look at setting up an alternative professional body, that's the way around it. But I don't have the energy or interest at all in doing it".

Unlike State of Grace, the FDANZ favoured professional embalming of the body "to ensure disinfection and preservation during the funeral period". Moreover, embalming was said to ensure the body "a more natural appearance" to those who wished to spend time with the deceased (FDANZ 2007c, p. 11). The Funerals New Zealand website, which appeared to be run by FDANZ, further promoted modern embalming as an essential service for the purposes of sanitation and preservation: "It enables everyone connected with the funeral – family, friends and professionals – to take part in rituals with no unpleasantness or embarrassment and without risk to their health whatever the cause of death" (Funerals New Zealand 2007c). With modern medicine working to prolong life, people were dying after their bodies had already started to decay. Moreover, with families dispersed around the globe, funerals were often delayed. Both these factors were used to argue for embalming to preserve the body in a pleasant and hygienic state for viewing until burial or cremation. Refrigeration was an alternative to embalming.

Living Legacies, which promoted more environmentally benign funeral alternatives, answered the question as to whether embalming was required as follows:

> A body has to be embalmed only in the presence of infectious disease. It also has to be disposed of before it becomes a public health hazard. If someone dies in a country other than the one in which they are to be buried or cremated, most airlines will insist the body be embalmed before it is flown home. In most circumstances of death a body does not have to be embalmed. (Living Legacies 2007a)

Regardless of the difference of opinion about the value of embalming, given the predominance of FDANZ-member conducted funerals, embalming was the norm in New Zealand, but the practice varied internationally. Besides New Zealand, embalming was common in the United States and Canada, but not as common in Europe and developing countries (Herald Journal 1997).

Deborah and Fran saw State of Grace's emphasis on not embalming as important to the sustainable business model they had in mind. They would accept clients where those clients or their families preferred embalming, but State of Grace did not encourage it because there was no suitable alternative to "formaldehyde and all the yucky chemicals", Deborah said. The business had an arrangement with a more traditional funeral directing firm to provide an embalming service – and with another to house the deceased until the funeral if the particular family did not favour keeping the body at home.

## Sustainable supply alternatives and acceptability challenges

State of Grace required a variety of generally available supplies along with some more specialist items. They used all natural products for cleaning the deceased. Natural soaps and shampoos were also readily available. The caskets State of Grace offered were all biodegradable and contained no plastic, beyond a necessary plastic liner beneath the body. To date, clients had not been particularly attracted to the more expensive, totally natural alternatives. A willow casket, made by special needs adults in a sheltered workshop where they grew and processed the willow on site, meant a lot to Deborah: "For me, it's perfect actually – it couldn't be a better product". At the other end of the price scale, the option of a cardboard box with a removable veneer cover could sometimes be perceived by clients as being too cheap. Getting completely sustainable caskets that were acceptable to clients was a challenge. According to Fran, "We're trying to find somebody that can do a plain, untreated pine casket for a reasonable cost. Good sustainable timber is what we really want to use".

Even the standard 'six feet under' approach to burials conflicted with the sustainability approach. "At six feet (1.8 metres), the earth is dead, you can't decompose, you're no good to anybody. It's clay. You have to be buried at three feet with a few spadesful of high competency compost mix thrown on top of the coffin", Deborah said. Deborah and Fran were working closely with Natural Burials and hoped that soon they would be able to offer a natural burial site in Auckland. That would mean being buried at three feet, un-embalmed in a biodegradable casket, and with trees planted on top. Both women believed natural burial sites would be much friendlier to the living than were traditional cemeteries.

But Deborah still worried that the business was not able to be completely sustainable. "The only area we've fallen down on is our transport, because we have to have a big car. We tried looking at a hybrid Toyota Previa, which is kind of a people mover, but no one really liked the look of that as a hearse". Instead, the business had a big Ford Falcon station wagon that was used only on the day of the actual funeral service. The trouble finding a suitable vehicle illustrated complications for a small start-up business accessing cost-effective, eco-friendly and socially acceptable supplies. For the two women, the world could not change fast enough in this direction.

## A wider trend toward the provision of more ecologically sustainable alternatives

State of Grace seemed to have tapped into a small but growing industry niche focussing on potentially more ecologically sustainable (or perhaps, more correctly, less ecologically unsustainable) alternatives to traditional means of dealing with the deceased. In some countries, remains could be buried in memorial nature parks where the overall setting was that of a wilderness rather than manicured lawn (Memorial Ecosystems 2007). Families could literally picnic nearby the often shallower graves where the composting body nourished the soil and supported plant growth.

Some of the statistics on the ecological impacts of traditional means of dealing with the deceased supported the position being taken by alternative firms like State of Grace, as well as the innovators of more radical techniques of disposing of bodies. Each year, the United States' 22,500 cemeteries buried: 827,060 gallons of embalming fluid; 1,636,000 tons of reinforced concrete (vaults);14,000 tons of steel (vaults); 90,272 tons of steel (caskets); 2,700 tons of copper and bronze (caskets); and 30-plus million board feet of hardwoods, much of it tropical (caskets) (Environmental Journalism Center 2002). Treated coffins took decades to decompose compared to untreated pine coffins, which tended to break down after 10 years (Bingham 2007). An embalmed body took about 60 years to decompose, with the possibility of leaching chemicals into soil and water (Bingham 2007). In the United States, the Green Burial Council maintained that unnecessary burial vaults put millions of tons of reinforced concrete into the ground and could leak toxins and heavy metals into the ground. Cremations, it claimed, used far fewer resources than did burials; however in an increasingly carbon conscious world, they could be criticised for burning fossil fuel. Mercury pollution was also emitted when a person with dental amalgam fillings was cremated, "though just how much is widely debated" (Green Burial Council 2007).

## Expressing social concerns in reclaiming death

Aside from seeking to reduce the environmental impacts of death, Deborah's and Fran's fundamental desire was to reclaim one of life's most important rites of passage. "I think a lot of people are told ... the more you spend, obviously the more you care, which is just ridiculous", Fran said. Hers was not a new concern. The Funeral Consumers Alliance, based in Vermont, was founded in 1963 to "protect a consumer's right to choose a meaningful, dignified, affordable funeral" (Funerals 2007). It argued that funerals often imposed considerable costs on families. Moreover, people were making decisions on what was required for a departed family member at a very vulnerable time, and often without knowledge of other alternatives available.

Instead, State of Grace wanted to give back to families their central role in death rituals. Its marketing brochure noted: "In times gone by, the family was responsible for the preparing of the body and other rituals associated with death. It was considered ... the definitive mark of respect and honour". The family could do as much or as little as they wanted to, but State of Grace encouraged people to be closely involved. Deborah and Fran believed having the family involved almost always helped people deal with grief. Generally, it also lowered costs.

Most people would not believe they could do what Deborah and Fran encouraged them to do. But they usually could. Deborah relayed the following example:

> We worked with a family with three siblings who hadn't really spoken for the last thirty years, and it was really uncomfortable when their mother died. We washed and dressed her. And things started happening very slowly, and by day two, it was like walking into a different space, they were holding hands around the table. Day three, there were hugs and tears, and this family was back together. And it was because they worked together for their Mum. It's just a

*matter of remembering that's what they've always done - looked after people they love.*

The two women objected to leaving death in the hands of people who equated professionalism with pretension. "Some in the industry think that what we're doing is unprofessional, because they like to present the body in such a way ... with a lot of ceremony, and tuxedos, morning suits, all that sort of thing", Deborah noted. She and Fran preferred a more personal and hopefully more cost-effective service – and the State of Grace partners took their social responsibilities seriously, wanting to be very up-front about both the realities of body care and the costs likely to be incurred in using their services.

## The economic cost of death

State of Grace was privately-owned, therefore, detailed company financial information was not publicly available. State of Grace had, however, made information on the cost of funerals available to the public. A typical State of Grace funeral cost NZ$3,000-4,000. More traditional funeral directors might charge NZ$4,000-$10,000 for a package for funeral-related services – but with 'extras' that cost could go considerably higher. Typically, costs included a service fee, the cost of a casket, transportation of the body, embalming, the burial plot or cremation, mourners' cars, newspaper notices, toll calls, medical and legal paperwork, flowers, organists, visitors' book, church, clergy, hall, catering and bar tab. Some people took care of some of these aspects themselves, but many seemed to prefer the funeral director to organise everything. Gravestones or urns were another cost which usually came later. Figure 10.1 provides an official FDANZ explanation of services that may be provided (FDANZ 2007c, p. 4). Table 10.1 compares State of Grace's advertised costs with typical industry averages (Prepaid Funerals New Zealand 2007).

**Figure 10.1** Services that may be provided by funeral directors

- Ascertaining the family's wishes

- Providing advice

- Transporting the deceased

- Liaising with the doctor, hospital or coroner as necessary

- Preparing, embalming and casketing the deceased

- Liaising with the florist, minister or celebrant

- Organising newspaper notices

- Organising the burial or cremation with local authorities

- Providing the authorities with the death certificate and burial information

*Source*: FDANZ 2007c, p. 4.

**Table 10.1** Average funeral costs in New Zealand compared with State of Grace[14]

| Services | Industry average cost range NZ$ | Average State of Grace costs NZ$ |
|---|---|---|
| Funeral directors' fees | $750-$3,000 | $1,150 |
| Casket | $500-5,000 | Starting at $350 |
| Burial including plot | $1,050-$3,150 | From $2,300 – varies considerably between cemeteries |
| Cremation | $200-$550 | From $260 – varies between crematoria |
| Plaque/headstone | $500-3,000 | Not listed |
| Service and/or Hymn Card | $100-300 | $3 for each service sheet |
| Newspaper notices | $100-$400 | $120 |
| Service donation | $75-$125 | $300 – varies considerably among celebrants |
| Flowers | $100-$300 | $200 for casket bar |

*Source*: Prepaid Funerals New Zealand 2007.

More widely, the funeral industry was seen as recession-proof, but without huge growth potential (Economist 2007). Even the big US-based public death care companies had generally not done well in recent times, having apparently paid too much in the scramble to buy independent funeral homes, and finding they could not wring big savings from them (Economist 2007). Another issue facing all competitors was the growing popularity of cremations, which were much less costly than were burials. In the US, average cremations cost less than half the typical burial – and the largest chain there was withdrawing from the cheapest cremations and going for more lucrative packages (Economist 2007). Another part of its strategy was to cut prices on caskets and urns and concentrate on selling bundles of services. The small New Zealand industry had not been immune to some of these wider trends.

## Local industry dynamics and competitors

Most New Zealand funeral homes were still owned by families or independent firms like State of Grace – and most offered a full range of services. In the 1990s, one of the big players in the US funeral industry, Stewart Enterprises, took over quite a few New Zealand funeral homes. "At the time, it was seen as part of the inevitable upsizing and multi-nationalisation of everything. But although the process worked

---

[14] In August 2007, 1 New Zealand dollar was worth 1.4 US dollars and 1.9 Euros (www.xe.com.ict 2007).

for textiles, motor vehicles, fast food and most other things, it failed with funeral homes", wrote one observer (Bone 2003). Stewart Enterprises New Zealand Limited sold its 22 funeral homes in August 2001 to the Australian-owned Bledisloe Group, which had retained a dozen or so dotted around the country. Notably, both acquiring firms operated the funeral homes with the former owners still in charge, and under the funeral homes' original names. The effect was argued by those in opposition "to preserve the comfortable family façade. Meanwhile the prices would go up. The original families would still run the show, but the profits had a new destination" (Bone 2003).

The 2007 Yellow Pages Directory listed 72 funeral directors in New Zealand's largest city, Auckland, including the 26 Auckland funeral directors listed on the FDANZ website Funerals New Zealand (2007e). Many appeared to be small firms, although a few had branches in more than one Auckland location. Through their advertising, some specifically targeted Maori, Pasifika or people with a religious affiliation. Others noted the presence of male and female funeral directors, the credentials of being a family firm of long standing as well as various service offerings including 24 hour service, service to all districts across Auckland, national and international transfers and their own chapels. Some directors noted that they worked in association with FDANZ firms.

For State of Grace, all 72 Auckland firms could be seen as competitors; however, a couple stood out as more direct competitors in terms of their Yellow Pages advertisements. A Touch of Heaven – Angel Funeral Services operated from South Auckland. Its advertisement boasted photographs of a hand and flower superimposed over an image of the sun setting over the sea – and led with: "We're passionate about the family we take care of…". Angel promoted a 24/7 service and a chapel facility with lots of parking. The Natural Funeral Company's advertisement showed sand ripples in circular formation and simply said "individually tailored care". Like State of Grace, The Natural Funeral Company's website gave a breakdown of likely costs involved, something uncommon in the industry.

Other natural funeral organisations in New Zealand included Living Legacies, based on the South Island in Motueka, and Natural Burials run from Wellington. Living Legacies claimed to "provide an environmental alternative to the existing funeral industry and a course of information and education about people's rights and responsibilities around death" (Living Legacies 2007b). Natural Burials was a not-for-profit organisation promoting the concept of natural cemeteries, involved in setting up sites in several centres. It had successfully lobbied for a natural cemetery in Wellington and was working to find a suitable location in West Auckland (Natural Burials 2007).

Although starting a funeral directing business could be relatively simple and cheap (the cost of a station wagon, a body removal tray or gurney and possibly somewhere to put the deceased until burial or cremation if not kept at home), established firms tended to have an advantage because attracting business depended very much on goodwill, reputation and recommendation. Apart from phone directory and limited press advertising, some funeral directing firms had relationships with particular rest-homes, left cards or brochures in hospitals, made themselves available to speak to groups of the elderly and/or offered tours of their facilities.

New Zealand funerals often occurred with a greater degree of informality than might be the case elsewhere. Maori funerals or tangi were more solemn with marae protocol more strictly observed (Immigration New Zealand 2007). During a tangi, the casket was left open for people to touch and speak to the body, as people believed the spirit did not leave the vicinity of the body until burial (Griefcare 2007b). Cremations were not favoured among Maori. A more culturally diverse population base in New Zealand and certainly in Auckland, where many immigrants settled and found work, meant there were opportunities to serve different communities' funeral needs in more culturally appropriate ways. Only about 5% of funerals in New Zealand were pre-planned by the deceased – considerably fewer than the 15% in Australia or more than 40% in the US (Funerals New Zealand 2007g). As a fledgling business, State of Grace had not actively encouraged pre-paid funerals, but the option was available through insurance companies.

## Aiming to make a living by making death sustainable

To survive that first year as an alternative start-up in an arguably saturated and tradition-focussed industry, Deborah and Fran had each put about NZ$15,000 into the business and not drawn anything from it. Any income went straight back in to buy supplies and pay for advertising. Advertising had been a large cost, which was ironic because they valued word-of-mouth referrals most highly. Nevertheless, brochures cost NZ$10,000 even though Deborah's brother did the work. The Yellow Pages advertising cost NZ$3,500.

Deborah and Fran did not see their unconventional approach to death as a challenge to the financial success of State of Grace. "I know that the time is right; it absolutely is right", Deborah said. She was more concerned about being able to meet the sustainability goals they had set for State of Grace. They were still trying to convince local councils that their cardboard caskets could be cremated without any undue emission problems. "Current cremation practices account for almost 1% of air pollution globally. We feel cremating an unembalmed body in a simple cardboard casket would be a huge improvement", Deborah said.

Fran worried more about the logistical side of the business, particularly about finding premises and upgrading the firm's vehicle. She estimated that to buy a commercial property in their area would be between NZ$150,000 and NZ$250,000. Renting or leasing premises would cost approximately NZ$15,000 a year, but they would need to set up the premises to suit the business and those set-up costs were unlikely to be recovered. Deborah and Fran also debated whether to lease or buy a vehicle. If they decided to buy, it would be a used vehicle, and Fran projected a maximum spend of NZ$20,000. All of these outlays meant more investment and ongoing expenses without a guaranteed income. Still, Fran hoped they could start drawing an income within another year.

# The rewards – and challenges – of a sustainable approach

In August 2007, State of Grace won the Supreme Award at the Northern Region's Sustainable Business Network Awards, and the company was set to compete in the national awards later that year.

The two business partners looked forward to having premises with everything on site, a cool-room in which to do body care, a place where families could stay the night if keeping the deceased at home was not appropriate, and space to allow for funeral services. The holistic centre could also provide work space for celebrants and grief counsellors – and be served by a full-time receptionist.

Deborah and Fran were both trying to cope with the strain of setting up the business and being on call 24 hours a day, seven days a week. In July 2007, both families went on vacation in Bali. The two women were not worried about the potential lost business. "We might not have one funeral for those two weeks", Fran said, "or we might have five and we've missed out on all of them. But you know, life goes on".

Until it doesn't.

# Epilogue

By November 2010, State of Grace had developed better than Deborah and Fran had expected, and the economic recession had not impacted their business. "People are still dying", Deborah explained. According to Deborah, revenues had "doubled every year and we now employ two contractors, have two hearses and draw a reasonable wage". Deborah estimated that 10–15% of customers chose State of Grace because of its sustainability focus. The business operated in leased premises with a viewing room, a space for showing caskets, a dressing area with storage, and a custom-made, walk-in cool room which could hold three or four bodies. Deborah admitted, however, that the cool room had already been at capacity on a several occasions and they were now looking to buy even bigger premises that would include their own chapel.

In 2010, there were no new industry entrants taking the same sustainability strategic positioning as State of Grace; rather, mainstream funeral homes were including 'green options' as part of their offerings. Tony Garing, Funeral Directors Association President, agreed demand for green funerals was growing and he was looking at setting up training programs for green funerals. "Funeral directors now offered recycled paper for service sheets, eco-friendly embalming chemicals and sustainable material for caskets, such as pressed carbon and cotton fibre", Tony said (Carville & Muir 2010).

Deborah and Fran did not see the growth of their business as undermining their sustainability values and principles. In fact, they found that the economy of scale meant some environmentally friendly supplies were getting easier to source. Deborah and Fran also felt that State of Grace's impact on changing the industry paradigm was becoming more pronounced.

Eco-coffin maker 'Return to Sender' was founded in 2007, partly inspired by Deborah's complaint to design friend, Greg Holdsworth, that there were no sustainable casket supply options. Textile artist and fashion designer Miranda Brown has designed a silk shroud, used instead of a casket. However in 2010, the most popular option selected by State of Grace customers was a cardboard lined wooden coffin. After the service, the cardboard liner is removed and the timber coffin is re-used, costing a third of the price of purchasing a coffin.

How much the industry would change was still unclear. In September 2010, Deborah and Fran's goals for the business were to continue to grow it steadily and to move into new premises with a chapel. Although they had met with a franchise consultant earlier in the year, they were unsure franchising was the right growth model and thought having branches in different cities might be a better fit. Deborah does not want to take it too quickly, "We're so busy right now that those kinds of longer-term plans remain on the back burner", she said.

# References

Bingham, E, 25 August 2007, 'The Green Dilemma', *Canvas Magazine, Weekend Herald*.

Bone, A, 25-31 October 2003, 'Profit from loss', *The New Zealand Listener*, Vol 19, No. 3301.

Carville, O & Muir J, 18 September 2010, 'Meeting Your Maker with a Green Conscience,' *The New Zealand Herald*.

Economist, 25 August 2007, 'America's Funeral Homes, Profiting from Loss'.

Environmental Journalism Center, 16 October 2002, *Tipsheet*, viewed 15 August 2007, <www.rtndf.org/resources/tipsheet/oct16.shtml#2>.

EverGreen Funerals 2007, EverGreen Funerals, viewed 6 July 2007, <www.livinglegacies.co.nz/evergreem.html>.

FDANZ 2007a, Burial or Cremation: Making a Choice [Brochure].

FDANZ 2007b, The Funeral Director [Brochure].

FDANZ 2007c, Funerals: Knowing What to Do When Someone Dies [Brochure].

Funerals 2007, viewed 15 August 2007, <www.funerals.org>.

Funerals New Zealand 2007a, Burial, viewed 6 July 2007, <www.funeralsnewzealand.co.nz>.

Funerals New Zealand 2007b, Cremation, viewed 6 July 2007, <www.funeralsnewzealand.co.nz>.

Funerals New Zealand 2007c, Embalming, viewed 6 July 2007, <www.uneralsnewzealand.co.nz>.

Funerals New Zealand 2007d, Caskets, viewed 6 July 2007, <www.funeralsnewzealand.co.nz>.

Funerals New Zealand 2007e, FDANZ Funeral Directors – Auckland, viewed 6 July 2007, <www.funeralsnewzealand.co.nz>.

Funerals New Zealand 2007f, Funerals No Longer What They Used to Be, viewed 6 July 2007, <www.funeralsnewzealand.co.nz>.

Funerals New Zealand 2007g, Planning Your Own Funeral, viewed 6 July 2007, <www.funeralsnewzealand.co.nz>.

Green Burial Council 2007, viewed 15 August 2007, <www.greenburialcouncil.org>.

Griefcare 2007a, What is Griefcare? viewed 6 July 2007, <www.griefcare.org.nz>.

Griefcare 2007b, Tangi – Maori Funeral Practice, viewed 6 July 2007, <www.griefcare.rg.nz/options/tangi.html>.

<http://www.herald-journal.com/Archives/1997/funeral.html>, viewed 14 November 2007.

Immigration New Zealand 2007, Funerals, viewed 6 July 2007, <www.immigration.govt.nz/migrant/settlementpack/onarrival/LifeAndLeisure/Funerals>.

Living Legacies 2007a, Frequently Asked Questions, viewed 1 August 2007, <www.livinglegacies.co.nz>.

Living Legacies 2007b, The Company, viewed 1 August 2007, <www.livinglegacies.co.nz>.

Memorial Ecosystems 2007, viewed 15 August, 2007, <www.memorial ecosystems.com>.

Natural Burials 2007, Natural Cemeteries in New Zealand, viewed 1 August 2007, <www.naturalburials.co.nz>.

Prepaid Funerals New Zealand 2007, Funeral Costs in New Zealand, viewed 4 October 2007, <www.prepaidfunerals.co.nz>.

Statistics New Zealand 2007, Births and Deaths March 2007 Quarter, viewed 1 August 2007, <www.stats.govt.nz>.

Sustainable Business Network 2008, About SBN, viewed 10 December 2008, <www.sustainable.org.nz/cms/index.php?page=about-sbn>.

Yellow™ 2007, Auckland [Yellow Pages Phone Directory], pp. 1151-1158.

<http://www.xe.com/ict>, Accessed in November 2007.

# Case 11

# Westfield talent management: Creating a high performance culture in Australian Operations Finance

DEXTER DUNPHY AND BRUCE PERROTT

## Introduction

This case documents the remarkable turnaround of the Australian Operations Finance Group in Westfield.

Over the 50 years of its history, Westfield has grown rapidly, establishing 119 shopping centres and expanding internationally into the US, New Zealand and the UK. With a history of separately listed Westfield companies facilitating that growth, a decision was taken in 2004 to merge the three entities – Westfield Holdings, Westfield Trust and Westfield America Trust – to create the Westfield Group.

The group became a global vehicle; the merger opened the possibility to realise the potential benefits of global scalability, cross-pollination of operational best practices and cross-regional efficiencies. The reality was that the current state of each region varied greatly in business systems and policies, and so a decision was taken to focus on and leverage best practice from the mature Australian business across the other countries – in relation to the core function of Finance in particular. In 2005 there were clear signs that the Australian Finance function required a significant revamp. It was not achieving the desired efficiencies or providing expected level of support to the operational business; this also hindered the progress of a global roadmap across the regions.

This is the story of how an organisation, over a period of four years, rebuilt the regional function into a high performance unit that has also become a net exporter of managerial talent to other parts of the company. The strengthening of the talent base and systems in Finance made an important contribution to Westfield's ability to maintain its financial performance through the global financial crisis from 2008 to the present, and the benefits are now well-progressed in being leveraged globally throughout the company.

# Westfield: the company

For over fifty years, the Westfield organisation has grown enormously in scale and complexity. However Westfield's core business has remained essentially unchanged, remaining focused on the development, design and construction of shopping centres, on operating and leasing them and on providing an attractive environment for retailers to trade and shoppers to shop.

Westfield has undertaken many evolutions throughout this time, relating to the corporate and financial structures, the management organisation, the geographical footprint, and the design and type of retail environment in the centres. These decisions have all been made with the intent of placing the group in an increasing position of strength to deliver on its core business objectives.

Westfield executives state that their competitive advantage is derived from the following factors:

- **a single business focus** on delivering premium shopping centre assets

- **vertical integration of the value chain** across property ownership, management, marketing and leasing through to property development, design, construction and funds management

- **diversification of the underlying rental stream.** The company currently has almost 24,000 retail outlets in four established, affluent markets

- **accumulated experience over 50 years in the shopping centre business.** Westfield has developed a strong, experienced management group with a distinctive style and culture. Continuing success depends on maintaining the experience and talent of the executive team and progressively building relevant competencies throughout the value chain

- **financial strength.** Westfield is now the world's largest listed retail property group by equity market capitalisation. The company's assets are valued at AU$60 billion with diversified global sources of capital. Each decade of the company's history reveals a story of growth in the scale of its operations and in returns for investors. Over the 50 years since listing, shareholders in the entities that today comprise the Westfield Group have contributed equity of AU$19.9 billion and received back some AU$15.6 billion in dividends and distributions. With a market capitalisation at 31 December 2009 of around AU$29 billion, some AU$25 billion of wealth has been created for Westfield shareholders

- **a global brand in both shopping centres and capital markets**. The Westfield brand itself is an important company symbol. When Westfield began business, shopping centres themselves were not branded. The W sign on the Westfield's centres is shorthand for Westfield's humble beginnings, in a 'field' in the 'west' of Sydney. The first red 'Westfield' sign was installed on the Burwood centre in Sydney in 1966. Westfield has also had a significant influence on the transformation of shopping centres from utilitarian buildings to large, upmarket shopping, eating, business and entertainment destinations

- **quality earnings.** These are derived from recurring income streams secured through lease contracts which are insulated from short term cyclical effects.

Throughout the years, strategies have principally been driven by a financially-driven philosophy, with operational efficiency complementing the financial outcomes, not necessarily as the principal driver.

This focus has resulted in a continuous review of the future, through the looking glass of long-term projections and forecasts, enabling proactive strategies to be actioned in order to meet Westfield's commitments to shareholders on growth and returns. Due to the type of industry Westfield operates in – which is characterised by stability, diversification of income streams and therefore a degree of predictability – the planning and forecasting process provides a high certainty of outcomes for the company. Looking ahead five and ten years, and anticipating and planning for potential change, has been a hallmark of the company.

With this financial focus at the forefront of strategic decision making, Westfield has always been prepared to change its structure where necessary, even when the existing structure has historically proven sound.

In the past decade, the most significant example of this was the decision in 2004 to merge the three Westfield entities – Westfield Holdings, Westfield Trust and Westfield America Trust – to create the Westfield Group. The rationale was to create a global operating and financial structure to match increasing international opportunities, and in so doing Westfield became the largest listed retail property group in the world by equity market capitalisation.

Financially, the significance of the merger was confirmed by a series of global transactions that followed its implementation, and the heightened capacity and strength Westfield achieved within global capital markets.

Operationally, the merger provided a significant opportunity and new dimension to roadmap for the business, with a global focus overlaying the established regional roadmaps.

Interaction across the operating countries had always occurred at an intellectual and cultural level; however, true integration and cross-regional operational efficiencies were constrained due to the management teams being quite separate.

With the merger, Westfield had an opportunity to develop a new roadmap, which encompassed a review of the potential benefits of global scalability, cross-pollination of operational best practices and cross-regional efficiencies as an integrated global company.

The reality was though, that the current state of each region was quite varied in business stage and life cycle. Each region, due to distinct business origins (organic growth in Australia, acquisition in the United States and New Zealand, and exponential transactional evolution in the United Kingdom), had to reach a level of reasonable commonality and standardisation prior to being able to truly leverage the benefits of being a globally integrated organisation.

Therefore, a decision was made to establish a common roadmap for the operations so that all regions were progressing in the same strategic direction, and to utilise the Australian operations – being the most established and mature part of the business – to explore and leverage best practice over time across the group. Ideally, a point would ultimately be reached where each of the regions could operate on a common platform of excellence.

This is all background to the rest of this case which centres on evolution and change in Westfield's Australian Finance function from 2005 to 2010.

## Transforming finance in Australia

As is often the case for organisations undergoing exponential growth, the commercial operations of Westfield grew faster than the sophistication of the 'back-office' functions. When Westfield's Australian operation was a smaller organisation, senior executives were able to meet and deal closely with all personnel. As the company expanded, the executive team recognised the need to redesign the core processes that supported the business.

The company was highly successful in building shopping centres and, in a rapid growth environment, the focus had been very much on property development. Finance was seen as a support function within the operational side of the business and had not demonstrated leadership in the Australian business. In addition, the support systems, including IT, were out-dated and had not matured with the business.

However, such enablers had become critical to ongoing expansion and to the flow of information from the operations to decision makers in the business. At a deep cultural level, Westfield takes pride in the micromanagement of information, which, in their view, facilitates thorough and commercially astute decision making. Within Finance though, in 2005, the requirement for the level of detail required was still achieved through highly manual means and staff-intensive processes. Systems investment had not been a high priority for capital investment, so a number of the systems were out-dated Enterprise Resource Planning (ERP) versions, which did not facilitate either the consolidation of information or its speedy availability for decision making.

The following issues had become apparent:

- Detailed financial planning and forecasting was highly developed but was built on complex mainly manual systems.
- Finance business support added insufficient value.
- The bulk of effort in Finance was directed at transactional processing.
- Processes were manual and labour intensive.

- Reporting timeframes were slow.

- There was a lack of integration between various financial systems.

- Finance aspired to become a business partner but had not achieved this.

## The case for change

At the end of 2005, staff turnover in Finance was excessive, reaching 50–60% per annum in some areas.

Finance teams were isolated and operating in silos. There was little opportunity for career progression so most people tended to leave after filling one role for a time. There was no succession plan for replacing those who left, and because of the ongoing high level of turnover, appointments were made with little reference to the future potential of those appointed. Some key positions remained unfilled for significant periods. Not surprisingly, staff engagement was low, and staff were working exceptionally long hours to meet the information demands of the business.

Despite a culture of delivery, staff in the Finance area struggled to meet required performance outcomes or engage effectively with the needs of the business. The GM Business Services commented that when he joined Westfield in 2006 "the IT systems hadn't been upgraded for ten years and people felt that they were the 'victims' of poor processes". The high turnover in the area was a major cost to the business and also affected the ability of the area to recruit good quality personnel as replacements. Amongst accountancy and finance professionals and recruitment firms in Australia, Westfield had a reputation as being a tough environment with poor developmental opportunity.

The National Accounts Payable Manager noted that three weeks into her new role at Westfield, she was asked for her views on how the area was operating. She said,

> We're not high performing, and it's not just a matter of poor communication. We are not invited to the table by the business. People are overworked. They're not using the technology or systems well and the systems are out-dated. There is no innovation and drive to find solutions as they are too busy processing transactions. After they've been here for fifteen months people say: 'I've done my time". They even say it's good to have a stint at Westfield on your CV to show you can handle the environment.

The high turnover and its associated costs were catalysts for launching a change program to conserve and develop talent instead of losing it.

## Instigating a change journey

By 2005 it was clear to Westfield senior managers that there was a cluster of significant problems in Finance that needed to be addressed. In addition, they saw the need to attract and retain the best people to maximise Finance's contribution to business performance. Clearly this was not just a set of technical systems problems; there was a perceived need to change both the technical systems base and the prevailing culture in Finance. To change the technical systems alone would have been counter-productive. It was also necessary to change the attitudes and values of those who work with the systems and make them effective.

As a result, the Westfield executive undertook to refresh the leadership team in Finance with experienced executives who brought a fresh approach to finance and systems strategies.

In 2005 and 2006, new appointments were made to Finance leadership positions. This new Finance leadership team then collaborated to define and push a strategy to build the efficiency and effectiveness of Finance and therefore, the decision-making agility of the business.

Rather than try to tackle Finance across the global group operations as a whole, it was decided to initially focus on creating "a centre of finance excellence (people, structure, processes, systems) in Australia, capable of being leveraged and replicated abroad".

This strategy included the following objectives:

- low cost transactional processing

- quicker reporting cycles

- single version of the truth for reporting

- more analysis and value add for the business in decision making

- integration between Finance and the business through valued commercial support.

A key principle of the strategy was to integrate the resulting projects, which focused on people, process, structure and systems. It was felt that to address one in the absence of integrating with the others would cause the strategy to fail.

A second key principle was to support the strategy with relevant expertise. As a result, staff were assigned to key project and change leadership roles, including a dedicated Human Resources Manager role (with a direct line to HR and dotted line to Finance) to focus on building and retaining talent.

Historically talent had been considered important at Westfield. The company prides itself on the fact that expertise and passion for 'going the extra mile' are core elements of the culture. Talent management has traditionally been considered important because of a belief that people are the real drivers of business growth and profitability. There is a clear rationale for this belief, which was articulated in interviews. This rationale took the following form:

- The market maintains confidence in Westfield through proven continuing high financial performance (in contrast to the majority of global competitors through the recent global financial crisis).

- Strategically placed new people bring in new ideas and innovative ways of doing things.

- The continuity of a talent-centred culture helps drive ongoing success and ensures a continuous transfer of intellectual property.

- Agility is maintained through the constant development of people and ideas.

Strong support from both Finance and the operational business leaders, including the Group and Deputy Group CFOs, Peter Allen and Mark Bloom, and the Managing Director of Australia and New Zealand, Bob Jordan, was also key to bringing about action to change the situation. So the Australia and New Zealand CFO, Paul Altschwager, was in a position to implement change with support from above. Early meetings of the key leaders concerned to change things in Finance resulted in a decision, in the words of the GM Leasing and Finance, "to invest in change, to drive through a new culture of performance, to focus on people and to create a structure that would support our future strategy".

Another key partner in the change team was Elizabeth Lovell, National HR Manager Corporate, who reported directly to the Director of Human Resources, Ian Cornell, with a dotted line relationship to the Finance leadership team. With specific intent within the strategy, she sat within Finance, embedded with the change team, for the first four years. The inclusion of a trained and experienced Human Resource Manager on the Steering Committee managing the change process played an important role in ensuring that the strategies for retaining, acquiring and developing talent were sound and effective. Elizabeth had an excellent background for the task with a strong academic training to Masters level in Finance, Marketing and Business Strategy as well as HR. She also had HR and industrial relations experience in the electricity, mining and building industries as well as extensive experience in coaching and mentoring.

It was not an obvious move for the Westfield executive team to incorporate a dedicated full time HR Manager into Finance because HR was only introduced into Westfield in the late 1980s and even then the degree to which HR was integrated into different divisions varied widely. What developed despite this was a process of renewal that was owned and driven by Australian Operations Finance and actively informed and supported by HR.

## Creating the change team and project teams

In order to achieve these objectives, the leadership team had to be invigorated with new appointments and provide new roles for existing talent. Those existing leaders who had talent were given redesigned or new roles and new leadership talent was sought from outside the company.

But leadership did not just remain at the top. Several integrated project teams were set up to work on problem issues that had been identified. Each project team had leaders who were chosen based on having a balance of behavioural competencies and technical knowledge. The purpose of these teams was not only to create change, but also to educate people about how to manage change and create a momentum for transformation. Project management expertise was introduced through external and internal training, coaching and mentoring and online project management systems. It was noted that Westfield was good at project managing buildings, but that formal project management in Finance was undeveloped by contrast.

## Early actions

The following early actions were taken to improve the situation:

### Drawing up a roadmap for talent management in Finance to optimise organisational effectiveness

The first task of the change team was to identify the cluster of issues affecting performance in Finance, to define what an effective talent management system would look like in the context of the broader Finance transformation and then to sketch out a roadmap to bring that system into effect.

It was also recognised that change was a journey that would occur over a period of time. External consultant resources were not used in the planning, and the strategies can be described as 'home-grown' using the combined expertise of the leaders, human resources and project teams.

Another key principle was that 'talent' applied to all levels, not just a chosen few near the top. The intent was to build a depth of investment in the pipeline, at both organisational and individual levels of capability.

There were four main elements of the strategy:

- *attracting and developing talent:* This included identifying existing and new talent, planning for future requirements and achieving a balance between internal and external recruitment. It also included finding ways to align the goals of the individuals with those of the organisation so that the area could develop a deep talent pool that could be used to respond to new challenges.

- *growing capability:* This involved growing both individual skills and organisational capability.

- *retaining talent:* This focused on building employee engagement using remuneration platforms, reward and recognition systems and the development of career opportunities.

- *maximising impact*: This involved building an environment in which talent at all levels could thrive and succeed through clarity of business strategy to all people in the division. Feedback would be tracked and leveraged on the function's performance by the business, leader behaviours and performance against a balanced scorecard, and organisational feedback through annual staff engagement surveys. Finally, the focus on increasing impact would involve an ongoing focus on the structure of the Finance group, which continued to evolve during the change process to become more embedded and closer to the business, with roles shifting to become less transactional and more commercial and value-adding.

To build impact, this meant leveraging the new foundation of good talent and its impact in generating value for the business.

The aim was to build a pipeline of staff with professional expertise and knowledge, able to demonstrate their ability to deliver operational excellence and who combine effective and efficient delivery with continuous improvement and evolution. These characteristics were seen as fitting in with the Westfield culture that stresses professionalism and resilience in the face of challenges. The aim was to put key people with a performance record and future potential in key positions and then use them to help identify solutions to the problems in the function.

## Building a new model of leadership for finance

One aim of making the cultural change was to develop a new cadre of leaders who could manage the function more effectively – some of whom would eventually provide leadership in other areas of the organisation.

The key elements of the new leadership model were seen to be:

- Thought leadership – having a long term vision, fostering innovation, acting strategically and building plans to ensure the delivery of short and long term results

- People leadership – being able to lead through others to achieve goals, building support and motivating others within their teams, leading people across functions as well as influencing upwards

- Results leadership – managing the execution of results and acting as a change agent for continuous improvement

- Self leadership – the ability to establish credibility and trust, to be authentic, to act as a role model for others and to coach and assess themselves so as to improve their leadership ability over time.

## New signals to the job market: the external recruitment agent meeting

One of the first actions of the change team was to invite all the external recruitment agencies to come together in one room and initiate a greater partnership with them. This was the beginning of an ongoing dialogue about the nature of the problems and how solutions could be found. Paul, Elizabeth and others outlined the problem as they saw it, invited comments and suggestions and described the 'roadmap' designed by the change team which they intended to pursue to turn the situation around. They asked for the cooperation of the recruiters in the process and committed to keeping them informed as the transformation developed. "We were a lot more open to our recruitment agents. We educated them around the change and the type of fit we were seeking. We gave them notice of roles that were coming up and all the while they were developing a better understanding of our business". The meeting has been repeated each year since to report on progress and invite feedback.

## Tracking and leveraging feedback to help drive transformation

In 2007, Westfield as a whole introduced a staff engagement survey. This provided a critical tool for tracking progress against key identified areas, celebrating successes and enabling best practice areas to be identified so other areas could tap into progress made.

Up until 2005, no engagement survey had been undertaken, and there had been a reluctance to ask people what they thought about the company. Introducing the survey was a signal that management was opening up to employees' views and was willing to take them seriously. "The surveys provide ongoing feedback. We work at getting people to accept these results and then creating action". The National HR Manager Corporate commented on the emphasis they place on returning survey results to participants in a timely way to ensure that the results are current. "The worst thing we could have done," she said, "would have been to have carried out the surveys and then ignored the results. This would have created a worse situation than we had".

# Further action: creating the transformational agenda

## Initiating key change initiatives to create a talent pool as the basis for future high performance

As a result of the first survey in 2007, four key areas of action were identified. These involved, first, major interventions in the Finance area centring on *increasing interdepartmental collaboration*, knowledge sharing and the visibility of leadership. Business forums and staff events were planned. The second action area was the introduction of measures to *attract quality staff*. This included measures to improve the brand image of Westfield Finance as an Employer of Choice and the introduction of improved tools and processes for recruitment. Measures were also taken to *retain quality staff*. Career planning was initiated and a common KPI (Key Performance Indicator) balanced scorecard implemented. Remuneration was benchmarked against the external market. Career information and tools were made available on the intraportal and a People Committee was set up to identify and deal with issues that were affecting workplace satisfaction. The fourth action area was an attempt to *eliminate wasted time and effort.* Investment was made in dedicated systems and process staff to provide support for improved performance. This involved making detailed studies of work and meeting efficiency, creation of a project management structure for internal projects and project teams improving business processes and reporting.

**Improving reward structures** A range of new initiatives were taken to create rewards for performance. Measures were taken to ensure that staff were competitively remunerated and had good working conditions. But this simply prevents dissatisfaction; what was more important was to create positive incentives. In addition, there was a focus on creating career development opportunities and encouraging lateral as well as vertical moves. In the past, a manager would simply 'tap someone on the shoulder' to fill a role. Now all jobs are advertised internally and everyone can apply. There is a focus on encouraging people to move throughout the organisation. In addition, some poor performers were also counselled out of the organisation. There is also a recognition system – the Westfield Way Awards – which brings achievements to the attention of everyone in the company.

Two years into the change process, the 2008 survey results demonstrated that these interventions had created traction and generated improvements. Results areas that had been relatively strong remained so, but other areas improved. For example, there was a 23% improvement in the perception that Westfield has a genuine interest in the well-being of staff. There was a 10% improvement in staff perception of the area's ability to attract and retain staff and an 11% improvement in perceived commitment to employees. Management was also seen to solicit ideas and opinions (up 10%).

There was, however, considerable room for further improvement. For instance, while the response to 'wasted time and effort' had improved by 9%, still 49% rated it negatively.

The subsequent 2009 results across Finance showed continued progress in attracting and retaining staff and in reducing wasted time and effort. Across most categories associated with the issues targeted in the ongoing change program, there was a steady upward trend. Clearly the initiatives taken by managers and working parties to address issues needing improvement were having an impact. It is not possible to point to one major intervention that worked magic; what this change program emphasises is the need for ongoing diagnosis, continuing workforce involvement in problem identification and problem solving, and managerial accountability for people processes as well as business results. It also demonstrates the value of system-wide, consistent and persistent effort to resolve problems and pursue opportunities.

Increasingly Information Technology was also being involved in the change program. There had been a lot of difficulties in IT system development, which caused risk to key enablers of the Finance strategies. There was a clear need for a co-evolution of the IT systems for the increasingly sophisticated role of Finance in the business. Joint teams were set up to integrate Finance and people objectives into the IT business plan.

'Linking roles' were created, referred to as Business Systems and Process Managers, who acted as key links in the communication process with responsibility for facilitating the system development process between IT and Finance. Part of their role was to challenge the existing and proposed processes and to work to ensure that the new processes were as simple as possible.

A clear example of the effectiveness of this can be seen in changes that took place in the forecasting process. When the changes began, forecasting was based on information from 10,000 tenants and took 25 days to assemble each month. With only five days remaining in each month, planning would start for the upcoming month straight away. After two years of developing the new system, it was still taking 25 days. So this was challenged. Why was so much time needed? Why was the process so complicated? What steps could be subtracted while preserving the integrity of the process? The time was cut down to three working days. "Before the changes that area had very high turnover, but over the next two years, only two people left".

Similarly, steps were taken to integrate Legal as well. A common business and HR agenda was being driven across all units within Corporate Westfield.

Increasingly the traditional high performance culture, which Westfield had always maintained but struggled to deliver in the face of manual processes, was enhanced through the transformation strategies. The objectives of acquiring and developing talent were working: major shifts had taken place in work/life balance, in corporate communications, in staff retention and people were highly engaged and committed. As performance improved, the function gained the respect of those in other divisions.

## Partnership with key stakeholders

Over time, with improvements in the Finance area showing results, emphasis moved to developing skills for interacting with, and influencing, key stakeholders. These included:

- external partners (joint venture partners and auditors)
- internal partners involved in the Westfield business operations .

The aim here was to play a more effective role in providing commercial insights into Westfield's decision-making process; to achieve a better balance between cost management and strategic advice on how to grow the business; and to contribute to governance, control and risk management.

This emphasis has had a significant effect on attitudes to Finance. "Meetings used to involve arguments about the numbers in the accounts. Now the discussion is more about the meaning and commercial implications of the information. Managers from outside the area have seen the value of getting information early, appreciate getting the right numbers and the focus on the real business issues. Finance has gained respect by being more transparent. So managers from other areas have moved from telling us how to do our job to communicating what they really want".

Over time the emphasis shifted from resolving internal Finance issues and building capability, to utilising that capability by creating crosscutting integrative structures linking Finance to its key client groups.

**Introducing 360 degree feedback:** Three-sixty degree feedback was introduced, with staff being expected to be knowledge- and people-leaders using their professional expertise. They were also expected to deliver excellent operational results and to combine effective and efficient operational delivery with ongoing improvement and evolution of systems.

But in Westfield Finance 360 degree feedback is only one of the many ways in which employee feedback is sought and used. 'Coffee chats' are a regular practice where a manager takes four or five people who are not from her or his own team out for coffee. They are usually from a range of different teams and they are encouraged to raise any issues of concern or discuss good practice. This information is fed back to other managers in Finance. There are also 'skip level interviews' where managers interview people in their own area one level below their direct reports. As a result of a range of methods for achieving upward feedback, staff feel that they can freely raise issues and that they will be listened to.

**The development planning cycle:** Development and career planning are separated from the management of performance objectives in order to reinforce the focus on development. The performance and development cycle operates throughout the year. It begins with objective setting which uses a 'balanced scorecard' approach stressing financial, people, operational and transformational/group objectives. Apart from meetings of managers with individual staff, there are people planning forums throughout the year where each leader is accountable to talk about their own people. These meetings are owned by the business with the process facilitated by HR.

# Successes and challenges

The results of this change program have been impressive and measurable:

- **Engagement** scores improved by 23% in 3 years.

- **Attractiveness to the market** improved significantly as shown by 20% of new hires coming through referrals and the costs of hiring were cut by more than half.

- **Career progression** improved with more than half the staff in Westfield Finance having two or more roles while at Westfield.

- **Succession planning** was put in place and there are now two or more successors in the pipeline for all key roles from manager to CFO.

- **Retention of staff** – 94% of high potential staff are now being retained.

- **Respect from the rest of the Westfield business** – in one year (from 2008 to 2009), the following changes took place in the response from other parts of the business:

  - overall satisfaction with Finance by the rest of the business increased by 11%

  - there was an increase in a favourable attitude to Finance decision support
    by 11%

  - satisfaction with the budget process rose by 11%

  - satisfaction with the forecasting process rose 25%.

Taken together these results can be seen as a 'vote of confidence' by both staff in Finance and in the rest of the organisation. The spin off for individuals was also apparent from the interviews. For example, the GM Leasing and Commercial Finance commented:

> The biggest learning for me is that I have learned about leading people and about the importance of people through the change process; how vital it is to get everybody on board with the vision. I have learned the importance of communication and of not moving too fast for the team I am leading. I regard myself as lucky to get to junior management level at my age and I still have a lot to learn, particularly in the technical area. I hadn't had 360 degree feedback before and getting it was helpful; in addition to that, I have had constructive feedback along the way. My biggest learning was about managing upward and demonstrating my leadership skills to those above me.

The GM Business Services noted the satisfaction he received from working in a very constructive interpersonal culture. He commented: "I like to see the staff growing and succeeding in their jobs. I get a real kick at seeing people overcome their weaknesses and build on their strengths". Comments such as these indicate that the emphasis on leadership development has paid off in terms of increasing confidence and competence on the part of managers.

The Australian Operations CFO described the overall approach to development as "seventy per cent on-the-job training, 20% coaching, networking and mentoring and 10% formal training – with four pillars that produce balanced managers: knowledge and experience, behaviours, technical skills and an emphasis on both culture and motivation". This is backed up by making leaders accountable for the people management side of their work. Twenty per cent of a manager's annual bonus rests on the retention of staff with potential, and on carrying out the full process of giving 360 degree feedback and ensuring that staff take leave.

## What lies ahead?

The change program successfully transformed the culture of Australian Operations Finance, earning respect for the division throughout the Australian operations of Westfield and producing measurable performance increases.

So where to from here? Paul Altschwager said:

> We have still a lot of work to do. Yes, we have had some big successes but we are still too dependent on a few people. The next challenge is to set up a simple structure. The current business plan has us restructuring the whole of Finance. We are agile enough now to move and to bed down structural changes and improved practices. The key to these changes is simplicity of operation.
>
> The future is about being increasingly relevant to our business. The question we should always ask is how we add value to our business in an efficient and effective means whilst balancing the desire to be business partners with a healthy dose of independence and control. We also need to continue to improve people's work/life balance and become even more effective in the way we support the business and add commercial value. Our recruitment emphasis will change too. We were recruiting at senior levels but we are now recruiting to junior roles, developing them and driving them up. What we want to do is

*create a highly motivating environment and this will bring Finance from being mainly a support function to being at the table as a full partner in the company's decision-making processes.*

The plan is to invest strongly in systems and processes in the future and to minimise paper and manual processes. The investment will cover significant changes to the company's core computer system and the changes will provide a leap forward into the latest computer technology. This will bring all the divisions into the same information system. Authorities are also being enlarged and "an example of this is that there used to be only a small number of people who had a corporate credit card and now 60% have one and can spend up to at least AU$1500 dollars on their own authority". The increase in freedom is accompanied by an integrated and detailed on-line reporting system and tighter controls.

## Leveraging value across the countries

Once a solid foundation in people, processes and systems had begun to emerge in the Australian business, work commenced to leverage the value generated across Finance into the other regions. In 2008, two inter-regional Finance Strategy conferences were held at an executive and operational level. The purpose of the sessions was to establish and agree on a global integration strategy for Finance and to develop a common roadmap highlighting opportunities for group standardisation, scalability, systems and people.

Outcomes of these conferences included the establishment of a Global Finance Operating Committee to govern the ongoing global transformation and ensure traction against the improvement roadmap, together with the formation of collaborative Global Process teams responsible for sharing and leveraging functional best practice, standardising data definitions across the group and driving aligned reporting, processes and systems.

These teams have resulted in the implementation of standardised data definitions across the group, establishment of a Global Data Warehouse and commencement of a standardised Business Intelligence (BI) platform and models across the group.

The deployment of the BI platform was a true example of identifying and leveraging best practice across the group. A world class budgeting and forecasting model and reporting methodology had been developed in Australia over a number of years. The opportunity to leverage this was identified and agreement reached to deploy this into the US and ultimately into the UK as a standard tool. This enabled the accelerated deployment of the models and reporting and resulted in the benefits realisation far sooner of adopting the established best practice than if each region had developed their result independently.

As progress on the global journey continues, many opportunities are being identified and leveraging across the group is becoming more frequent and stronger. The value in group wide collaboration and standardisation is becoming increasingly recognised and explored amongst the countries.

# Key lessons from the case

We were particularly intrigued by this case because of the use of transformational change management strategies and systematic human resource management tools within the finance area of a firm. In our past experience, many finance functions have been slow to see the advantages of these approaches and reluctant to support and resource them in other areas of the organisation, let alone in their own. There is an important lesson here therefore for professional finance managers – even in their own area, these approaches can substantially improve performance and help create an environment that attracts scarce first class talent.

However we believe that the basic approach described here is applicable to organisations generally. As we move into a knowledge-based society, attraction and retention of talent is the most important ingredient in corporate success. Without outstanding people and a constructive culture, the best business strategies are ineffective. People make strategy happen. Modern organisations require engaged and committed employees at all levels who are partners in business development, operate efficiently, add value to the firm's operations and who are committed to continuing innovation.

A key differentiator was also the fact that the business and human resources maintained an integrated approach to change, and operated in clear partnership, with ownership of the results and strategies residing with the business, rather than an external HR or consulting expertise. This has greatly enabled people management and change expertise to become embedded with the finance talent, increasing capability for ongoing transformation within the pipeline of leaders and successors for the people strategies.

IBM's 2008 Global CEO Study *The Enterprise of the Future* (1) asked more than 1000 CEOs of large global companies about the future they were facing. The large majority saw that they were facing substantial or very substantial change over the next three years. This was before the GFC – few would have anticipated the extent of the change that was to occur. But even so, most also reported that their organisations were experiencing increasing difficulty in managing existing levels of change effectively.

In a subsequent IBM study directed to investigating this further, *Making Change Work* (2), it was confirmed that most CEOs consider themselves and their organisations to be executing change poorly. This latter study interviewed 1500 managers of large-scale change projects and found that only 41% of these managers themselves, each of whom had direct responsibility for a project, considered the project "successful in meeting project objectives within planned time, budget and quality constraints. The remaining 51% missed at least one objective or failed entirely" (p. 2). But the best 20% of these organisations, referred to in this study as 'Change Masters', "reported an 80% success rate, nearly double the average" (p. 2) and ten times better than the bottom 20% of the sample.

Clearly some organisations manage to build in an ongoing capability to manage transformational change. In an increasingly turbulent environment, those organisations that succeed in building the skills for managing change throughout the organisation are those that will survive and thrive in increasingly turbulent environments. Building human capability is an important aspect of creating more sustainable high performance organisations.

In our view there are lessons to be learned from this case by all managers who are concerned to create resilient, high-performing organisations ready to shape the emerging future.

*Acknowledgement*: We appreciate the willingness of Westfield to allow us unrestricted access to people and information needed to write this case.

## References

IBM Corporation 2008, *The Enterprise of the Future*, Global CEO Study, IBM Global Business Services, NY.

IBM Corporation 2008, *Making Change Work*, IBM Global Business Services, NY.

For further information, researchers and students can access <westfield.com.au>.

# Case 12

# Westpac Banking Corporation: What do we mean by sustainability?

Tamsin Angus-Leppan and Tim Williams

## Introduction

During the early 1990s in Australia there was a public outcry against banks closing their rural branches and charging what were perceived to be exorbitant fees. Bank staff were targeted by angry customers and as staff morale plummeted, the major Australian bank Westpac realised it needed to fundamentally examine and respond to what seemed to be intractable reputational challenges plaguing the banking industry. This led to a more structured approach in an attempt to: understand shifting community expectations; measure and report more broadly on Westpac's impacts in the community; and clarify social, environmental and customer commitments – what the organisation stands for and what stakeholders can expect.

Westpac built on a legacy of social responsibility with initiatives like a new board committee focused on sustainability, structured stakeholder engagement, and public reporting using frameworks like the Global Reporting Initiative. Westpac's progress was recognised with awards, and the company topped the Dow Jones Sustainability Index (DJSI) in the bank category. However, some of the bank's stakeholders are sceptical of these developments and hold diverse perceptions of what corporate sustainability is and should be at Westpac. This case explores the challenges in gaining shared understanding of corporate sustainability amongst employees, customers and other stakeholders and how this can impact organisational change towards sustainability.

## Background

'Corporate sustainability' is an ambiguous term; there is considerable debate about the range of organisational features (van Marrewijk 2003) and wider socio-political factors (Matten & Moon 2008) it includes. Corporate sustainability is most frequently understood as a broad prescription for company responsibility to *stakeholders* – i.e. those parties who influence and are influenced by the activities of the firm – including but not limited to shareholders (Freeman 1984). It can be thought of as overlapping with corporate social responsibility and as having three basic elements: economic, human and ecological sustainability (often interpreted as the 'Triple Bottom Line' of economic prosperity, environmental quality and social justice (Elkington 1998)).

Westpac is rated highly for corporate sustainability as judged by independent assessments like the Dow Jones Sustainability Index. This case focuses on a range of stakeholders of the bank's corporate sustainability initiatives, which are wide-ranging, extensively reported and supported at board level. The initiatives include:

- leadership on panels lobbying on climate change policy

- serving as founding members of international initiatives like the United Nations Environment Program and Equator Principles[15]

- initiating highly-regarded community partnerships

- Initiating award-winning employee volunteering schemes

- pioneering sustainable supply chain management

- upholding principles for responsible lending

- the continuing challenge of sustainability-related products and services.

## Methodology

The stakeholders interviewed for this research are employees, consumers, suppliers, NGOs that form part of Westpac's Community Consultative Council, and Landcare[16], a community partner to the bank. We selected a cross-section of bank employees whose job functions involve them in sustainability issues at least peripherally and/or who had indicated commitment to ecological and/or human sustainability issues. The final sample consisted of 85 informants representing 21 in-depth interviews and ten focus groups. The research was of an open-ended exploratory nature, asking how stakeholders perceive corporate sustainability. As there is debate about the definition of corporate sustainability, the meaning of corporate sustainability was left open and explored with research participants.

---

[15] Signatories to the Equator Principles agree to provide loans only to those projects whose sponsors can demonstrate their ability and willingness to comply with processes that ensure that projects are developed in a socially responsible manner, according to sound environmental management practices.

[16] Landcare is an Australia-wide network of community groups who practise sustainable land use and regeneration.

# Analysis

The interviews and focus groups were transcribed and analysed using content analysis software called Leximancer. Leximancer uses machine learning to determine the key concepts in the text, and the relationships between them. A concept is a set of words that travel together through the text.

# Results

In Leximancer, the frequency of co-occurring concepts is measured, weighted and clustered to produce a two-dimensional map of concepts (for further details of this process see www.leximancer.com). The Leximancer map is a snapshot of the cognitive structure and content of the data at the macro level indicating:

- Concept frequency: the concept labels in the map shown in Appendix 12.1 (e.g. 'bank') range in colour from black to light grey. The darker the concept's label, the more frequently it occurs in the text.

- Contextual similarity: Adjacency indicates that two concepts appear in similar conceptual contexts.

- Similarity of views between informants: Tag classes representing informant groups (e.g. 'TG_CONSUMERS_TG' in Appendix 12.1) are positioned around the edges of the map. Their proximity shows the similarity of views and the concepts they have strongest association with.

Leximancer also produces a ranked concept list (see Appendix 12.2) showing the relative frequency with which concepts were discussed in the text.

# Diverse perspectives on corporate sustainability

Two important outcomes from the research were that:

1.  In terms of similarity of views amongst informants, it appeared that the text could be broken into eight distinct groups: executives, middle managers, branch staff, contact centre staff, community partners, suppliers, consumers and the Community Consultative Council – each indicated on the map by a tag label. The fact that these tags are spatially separated on the map indicates that there was little in common between the eight groups, hence we concluded that there were divergent perspectives on corporate sustainability across Westpac's stakeholder population.

2.  Stakeholders' understanding of corporate sustainability at Westpac was that it was comprised of five human and ecological elements: green products, environment, social, work and community. Looking to the ranked concept list in Appendix 12.2, the list shows 'human' concepts: 'social', 'work' and 'community', as the top three concepts. Close behind are two concepts representing ecological elements: 'environment' and 'green (products)'. The remaining human and ecological concepts on the list have been combined; we have absorbed 'environmental ' into the

'environment ' element and 'working ', 'job ', 'staff ', 'worked ' and 'leadership ' into the 'work ' element.

However, there was considerable divergence in opinion between the groups concerning a) whether and how each element contributes value to the bank and b) whether and how each element is embedded as part of the bank's policies and procedures.

Table 12.1 illustrates the divergence in opinion between the eight groups with examples of quotes taken from the interviews and focus groups with each stakeholder group.

Table 12.1 indicates the divergence in opinion between stakeholders as to what corporate sustainability is and should be at Westpac. Overall, the observed differences between stakeholder perspectives on the corporate sustainability elements belies the concept of corporate sustainability as a 'shared value' driven by consensus in organisations (Waddock & Bodwell 2007). Note, for example, the Environment quotes in Table 12.1. Middle managers tended to emphasise the environment as a popular issue with comments such as, "The sustainability issue of the day is the environment, not social issues"; whereas branch staff saw environmental issues as not relevant to the bank and consumers and indicated that they did not see environment as a popular with the public in general: "I'm not very environmentally conscious because society isn't really pushing that".

# Implicit and explicit corporate sustainability

A further interesting outcome of the research was that there was support for a view of corporate sustainability as being implicit or explicit, reflecting two competing approaches to CSR (Matten & Moon 2008). Implicit corporate sustainability is represented by "values, norms and rules, that result in (mandatory and customary) requirements for corporations to address stakeholder issues" (Matten & Moon 2008, p. 409). Explicit corporate sustainability, on the other hand, is voluntary and implemented as a result of deliberate and often strategic decisions made by the corporation. While the corporate sustainability practices might be the same for firms practising implicit versus explicit corporate sustainability, Matten and Moon (2008) emphasise differences in language and intention. Corporations practising implicit corporate sustainability view it as an implicit element of the institutional framework of the corporation and do not describe values or norms-driven practices as corporate sustainability; whereas those practising explicit corporate sustainability communicate their policies and practices to stakeholders using the language of corporate sustainability as a strategy to foster shared meaning of its business value.

**Table 12.1** Exemplary quotes: the elements of corporate sustainability according to Westpac stakeholders

| Stakeholder groups | Green products | Environment | Social | Work | Community |
|---|---|---|---|---|---|
| Middle management | (We need) an infrastructure that makes all our products green | The sustainability issue of the day is the environment, not social issues | There is almost a fear of polarising the group | I tell people I work in CSR, which is not like working for a bank | Some people would see this (community involvement) as fluffy |
| Consumers | Green shares are not necessarily good for the bottom line, which is what we get paid from. | I'm not very environmentally conscious because society isn't really pushing that | Interest always makes rich people richer and poor people poorer | We're not going to read an article about a boss who's nice to his employees but we'll read about sweatshops | Corporations don't have to advertise it to make money out of it; they should just do it |
| Community partner | Green products don't shine out at me in the supermarket and I work in that area | Landcare is not about the diversity of species, it's more about leaving it to your children in a better shape than when you got it. | There is no environmental problem, there's a social problem | I'm happy working with companies as long as they are making progress towards sustainability | A number of people would say they're just trying to clean up their name by borrowing our reputation |
| Executives | How can we (the bank) have a green bank account? I don't have the imagination to take you there | Changing behaviour around climate change is about getting people to consider a wider set of value drivers | It's more that people will say 'it's against our values' than they'll kind of tie it to social responsibilities | 80% of staff cite the CSR as the reason they started here but it doesn't help staff retention | Our community work wins awards |
| Branch | We've never had anyone ask for a green product | I don't think people associate banks with environmental issues | On the social side, things that could go wrong we are sorting out | Country staff are your most loyal staff | We do a lot of voluntary work but I'd hate to think we only do it for the publicity |

| Stakeholder groups | Green products | Environment | Social | Work | Community |
|---|---|---|---|---|---|
| **Contact centre** | Don't know much about green products | What I'd like to see the bank do is become completely carbon neutral | A lot of people are negative towards banks and we can all understand why | (Head Office) should make it known to staff what they actually do | I just use our community work as some ammunition to hose people down a bit |
| **Suppliers** | You have to be careful that, whilst you have real green products, your other ones have some element of responsibility too. | You become cynical when a company says we're not going to do paper statements because it's environmentally unfriendly | I spend quite a lot of time writing and helping banks create direct marketing packs to get someone to take out more debt | I think potential employees are doing a lot more questioning about corporate values | When a company becomes over involved in one of those neutral companies then there's a suspicion of it |
| **Community consultative council** | I don't get the sense there's the culture (at the bank) that is necessary to make (green products) really work | | What I'm concerned with is what your core business is doing and what impact that's having on society | There are people who are absolutely passionate about corporate sustainability at the bank | That stuff (with aboriginal communities…I thought that was great and I hope you can continue to do it |

In this study, the executive group seemed to favour implicit forms of corporate sustainability, making comments (cited in Table 12.1) such as, "It's more that people will say 'it's against our values' than they'll kind of tie it to social responsibilities". Also, the research showed a backlash emerging against explicit corporate sustainability. Looking, for example, at the Community column in Table 12.1, the comment from consumers that, "Corporations don't have to advertise it to make money out of it; they should just do it", indicates a sense from consumers that community initiatives should be implicit, as does the comment from the branch group that, "We do a lot of voluntary work but I'd hate to think we only do it for the publicity".

## Conclusion

This case explores the potential for an understanding of corporate sustainability that is shared across the organisation. By examining a range of stakeholder perceptions of Westpac's corporate sustainability, the case suggests that, even in our leading sustainability organisations, assumptions cannot be made that corporate sustainability is enacted as a shared value. Not only is it the case that views may not be shared, they are likely to be in conflict. Other research indicates that this is not an uncommon problem in organisations. This begs the questions: what is it about sustainability that makes this shared understanding so difficult to attain; is it important for continued organisational commitment to sustainability; and how can companies best communicate to customers on sustainability?

# Appendix 12.1

Leximancer map of data from middle managers, branch, executives, suppliers, contact centre, consumers, community consultative council and community partner

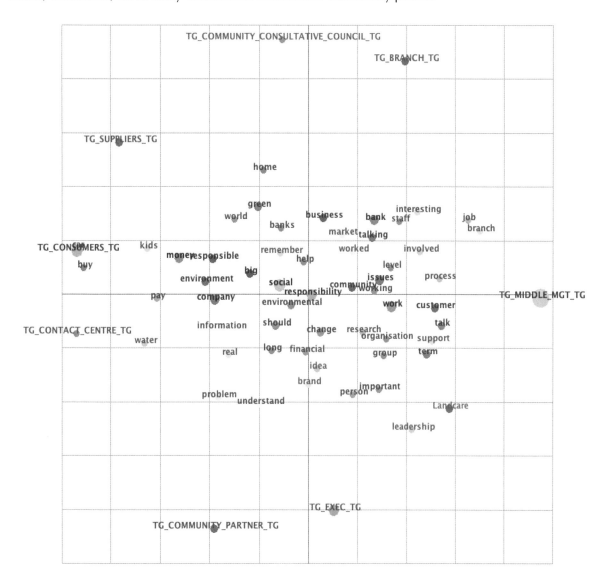

# Appendix 12.2

Partial ranked concept list (contains concepts relevant to this analysis only)

| Concept | Relative count |
|---|---|
| social | 17.2% |
| work | 14.5% |
| community | 7.5% |
| environment | 6.4% |
| green (products) | 6.1% |
| environmental | 4.2% |
| working | 4.1% |
| job | 3.7% |
| staff | 3.6% |
| worked | 2.5% |
| leadership | 2.2% |

# References

Elkington J 1998, *Cannibals with Forks: the Triple Bottom Line of 21st Century Business*. New Society: Vancouver.

Freeman RE 1984, *Strategic Management: A Stakeholder Approach*. Pitman: Boston.

Matten D & Moon J 2008, 'Implicit and explicit CSR: A conceptual framework for a comparative understanding of Corporate Social Responsibility'. *Academy of Management Review* 33(2): 404-424.

van Marrewijk M 2003, 'Concepts and Definitions of CSR and Corporate Sustainability: Between Agency and Communion'. *Journal of Business Ethics* 44: 121-132

Waddock S, & Bodwell C 2007, *Total Responsibility Management: The Manual*. Greenleaf Publishing: Sheffield, UK.

# Case 13

## Yarra Valley Water: Learning and change for sustainability

PATRICK CRITTENDEN, SUZANNE BENN AND DEXTER DUNPHY

## Introduction

Yarra Valley Water (YVW) is widely recognised in Australia as a leader in corporate sustainability. The Victorian Government-owned water utility delivers water and sewerage services to over 1.6 million people in the northern and eastern suburbs of Melbourne. Its operating licence covers over 4000 square kilometres and it maintains a distribution network comprising over 16,000 kilometres of water and sewerage pipes. Operational challenges include the maintenance of ageing water and sewage infrastructure in established areas and the development of new infrastructure in the rapidly expanding northern suburbs of Melbourne.

*Challenge*

The organisation's capability and commitment to delivering sustainability outcomes has been demonstrated at a practical level through the implementation of innovative projects. Its contribution as a leader has been recognised through public sustainability awards at state, national and international levels. Stand-out sustainability projects include the design and management of

- one of the largest recycled water systems in Australia at the Aurora residential development

- a pressure sewer system in the outer Melbourne suburb of Gembrook, which delivers an improved environmental outcome in compared to traditional gravity-based sewer systems through reduced energy use and having pipes laid at shallower depths

- a stormwater filtering and reuse project that is currently underway in a new industrial development at Kalkallo.

YVW's ability to consistently deliver innovative projects and to demonstrate sustainability leadership in the water industry is the outcome of a decade-long focus on organisational learning and change. This case study describes three inter-related aspects of YVW's approach:

1. **Organisational culture.** Since 2001, YVW have been implementing management initiatives to create a more open and collaborative organisational culture that has fostered innovation and creativity.

2. **Integration of *environment* as a strategic issue.** In 2003 YVW established 'environment' as one of four key elements of the organisation's strategic intent and continue to integrate environmental considerations into core business decisions.

3. **Sustainability tools and approaches.** Since 2003, YVW have developed and integrated a number of sustainability tools and approaches, including The Natural Step, Life Cycle Assessment and stakeholder consultation, to support more effective decision-making at all levels of the organisation.

These three central aspects of the YVW approach have worked together to mutually reinforce ongoing and effective learning that has resulted in impressive business and sustainability outcomes. The three aspects are graphically demonstrated in Figure 13.1 and described throughout the case study.

**Figure 13.1** Interrelated aspects of organisational change for sustainability at Yarra Valley Water

# Methodology

This case study is based on face-to-face interviews with eleven senior managers; the interviews were conducted in September 2009 by researchers from the University of Technology Sydney and ARIES, Macquarie University[17]. Additional information was obtained in November 2010 to update the case study. Each interview was approximately 45 minutes in length. The interviews were transcribed and a qualitative analysis was undertaken to identify key themes. Publicly available case study material and presentations that have been made by YVW managers were also reviewed and incorporated into the analysis.

# Background

Yarra Valley Water was established in 1995 at the start of a period of rationalisation in the Victorian water industry. At that time, three water retailers were established and allocated exclusive operating areas in Melbourne. Historically, YVW was part of Melbourne Water (1992–1995) and before that (1890–1992), the Melbourne Metropolitan Board of Works. The Victorian Government's intent in creating YVW and the other water retailers was to introduce competitive elements that would improve efficiency and service. At the time, the government was also working through a process of privatising energy utilities. A change of government in 1999 diminished the possibility that water would be privatised and it remains a government-owned corporation.

The 'comparative competition' regime introduced by the government helped drive a customer service culture and improved business management systems. For example, in 1996 the company's environmental management system was accredited under the ISO 14,001 environmental standard.

# Approaches to learning and change at YVW

## Creating a more innovative organisational culture

Organisational culture is often defined in simple terms as 'the way we do things around here'. In 2001, there was a sense at YVW that to be successful, the organisation needed to be more innovative in its approach. Developing a better understanding of the organisational culture, and acting to improve it, was considered an important step in supporting greater innovation.

The cultural change program began with the management team completing an organisational culture questionnaire, developed by an international consulting company, to determine the *preferred* culture for the organisation.

---

[17] The project was funded by the Australian Government Department of Environment, Water, Heritage and the Arts.

*Our goal was to build a high performance culture in which managers display constructive behaviours that facilitate high-quality problem solving and decision making, teamwork, productivity, and long-term effectiveness.* (HR Manager quoted in Jones *et al.* 2006, p. 194)

Management and employees then completed a second organisational culture questionnaire to determine the *actual* operating culture. The results highlighted that the *actual culture* was characterised as *defensive, avoidance-oriented, oppositional and competitive*. It was summed up by one senior manager in the following way:

*We were very competitive internally. The culture of the place was that you would set one group up against the other competing for budget in a very overt way, it was win /lose. So we were competing with each other, rather than with the water industry; so that was really unproductive.* (MD, quoted in Jones *et al.* p. 186)

This hierarchical and somewhat bureaucratic environment, with fairly authoritarian leadership, drove workplace practices such as:

- avoiding responsibility and action by delegating upwards

- avoiding and blocking new ideas

- using policy and procedures as a way of limiting change and innovation.

Perceiving the gap between the preferred and actual operating culture was somewhat disturbing for both management and staff. However, open communication of the survey results ensured that the challenges were not hidden and staff were encouraged to take responsibility for contributing to the development of a more constructive workplace culture.

Selected management initiatives were implemented to bridge the gap between the preferred and actual operating culture.

These included:

- development of an agreed set of organisational values

- active encouragement of problem solving and project development through cross-functional work teams

- introduction of skip-level interviews which involved employees speaking with their manager's manager on a regular basis

- monthly meetings of staff from all levels with the Managing Director – all staff can volunteer to participate, with five or six randomly selected each month

- review of reward and recognition systems to ensure preferred cultural behaviours are encouraged

- annual 'Blue Zone days' in which preferred cultural behaviours are explored and clarified through fun and engaging activities. Blue Zone" refers to the constructive styles on the Organisational Culture Inventory (Jones *et al.* 2006), which defined preferred cultural attributes.

Cultural surveys are now conducted every two years and there has been a steady improvement towards a more open, collaborative and constructive organisational culture.

## Integrating *environment* as a strategic issue

In 2003, following the appointment of a new CEO, Tony Kelly, a series of collaborative staff workshops were held to develop a clearer sense of direction for YVW. Work on changing the culture had continued, but a clear and agreed direction for the organisation was essential to focus staff effort. The changes already implemented in the culture meant that this sense of shared direction was much easier to develop than it would have been in the earlier period, characterised by intra-organisation competition.

> The strategic intent and aims were established as follows:
>
> Yarra Valley Water's strategic intent is to lead the global water industry in serving the **customer** and the **environment**, supported by our high performing business **culture,** and continuously improving our **efficiency**.

By 2013, Yarra Valley Water aims to achieve the following outcomes:

- **Customers** – Our customers recognise us as their best service provider and are engaged in what we advocate.

- **Environment** – We provide services within the carrying capacity of nature and inspire others to do the same.

- **Culture** – We have a vibrant workplace achieving exceptional business outcomes, successful partnerships and personal satisfaction.

- **Efficiency** – We achieve our objectives at the lowest community cost and consistently met our shareholders expectations.

These strategic objectives are broken down into functional activities. Targets for each functional activity are set and monitored monthly. The executive team report progress to the board on a quarterly basis. The strategic objectives have been maintained over the past eight years and at the time of writing progress towards the outcomes continued largely as planned. There has been some modification of supporting objectives, but the strategic objectives have proven to be resilient.

There were a number of important business drivers that led to the incorporation of environment as a strategic business issue.

These included:

- the lengthy drought that began in 1999

- community tensions over water planning and selection of 'right' supply options

- changing expectations amongst customers regarding water savings and efficiency

- recognition of the potential environmental impacts resulting from the company's operations and the need to contribute effectively to global issues such as climate change.

By incorporating environmental responsibility as an explicit strategic objective, initiatives to improve environmental outcomes were actively encouraged. Environment became an important and legitimate focus for staff at YVW. It was decided that integration of environmental considerations should be encouraged wherever possible.

A turning point for Yarra Valley Water was learning that environmental focus and business success were mutually linked. An environmental education program was used to inform all employees of the importance of the environment, while exercises with the management team that analysed past, present, and future projects made it transparent that projects could be selected that delivered both environmental and business value – and those were the ones that should be pursued.

> *We started off with a very conscious decision that we will not have an environment department, or one person doing it, so it's integrated into everyone's job.* (Manager, Research and Innovation)

Other ways in which environment was integrated into management practices included establishing specific targets and incorporating environmental responsibilities within job descriptions and performance management.

The bar had been set high, and the challenge was to deliver results that demonstrated *leadership in the global water industry*. A consultant working with the management team at this early stage expressed the challenge in the following way to highlight the potential negative impact that a decision can have on an organisation's culture:

> *Isn't life frustrating if you're always being hit over the head by the regulator? Isn't it more exciting for an organisation to be one step ahead of the regulator, and if you can see what's happening – then create your business in that space?*

It was recognised that conventional environmental management approaches of just setting goals would not be sufficient to achieve the goal of global leadership and recognition and management began the task of identifying and developing the tools and approaches that would be help YVW achieve this goal.

## Sustainability tools

Turning a strategic priority into practical initiatives saw YVW engage in organisational learning supported by a range of specific sustainability tools and mechanisms.

### The Natural Step

The first sustainability-oriented tool that was applied was The Natural Step (TNS). The objective of TNS is to create a shared understanding and language for sustainable development with the intention that this will, in turn, lead to improved sustainability outcomes. At its core are four scientifically based system principles:

*In the sustainable society, nature is not subject to systematically increasing…*

1. *Concentrations of substances extracted from the Earth's crust.*

2. *Concentrations of substances produced by society.*

3. *Degradation by physical means.*

4. *And, in that society human needs are met worldwide.* (Robèrt *et al.* 2002)

A program was developed across YVW to build an understanding of the systems principles and to use them as a basis for decision-making.

One senior manager summed up the major benefits of TNS as:

*We used The Natural Step to help us understand what sustainability was. Back in 2002, sustainability was a vague concept. The scientific principles were particularly useful to us, as we are largely an engineering organisation, so the scientific language worked well. The Natural Step team at the time also had some very competent senior strategic thinkers that could engage our Executive. Their contribution and in particular their ability to show us how we could improve in environmental sustainability AND improve business value should not be underestimated.* (See also Pamminger & Crawford 2006)

Amongst the management team, however, there was some frustration that this approach was not leading to practical outcomes and was limited in its strategic application. MD Tony Kelly, expressed this as:

*Where we struggled with TNS is that it really didn't help us work out what we had to do on Monday. They gave us the beacon on the hill which was the thing to aim for, which was great and the principles are very sound I think, but after 18 months an unanswered question for us was, "What are we going to do tomorrow?"*

## Life Cycle Assessment

Following the introduction of TNS, Life Cycle Assessment (LCA) was trialled on a number of projects including the analysis of recycled water options and the environmental considerations associated with the use of water tanks. The aim of LCA is to take a whole-systems approach to identify and evaluate the environmental impacts of products and services from raw materials extraction and processing through to end of life.

By working across a more extensive value chain, opportunities to identify and act to reduce environmental impacts are enhanced. This approach contrasts with traditional environmental assessment processes that typically focus on the operations more directly in the control of one particular organisation. Traditional approaches such as these may eliminate significant opportunities to reduce environmental impact that may only be achieved through collaboration of the various stakeholders that make up that value chain. LCA avoids shifting problems from one organisation, geographical site or life cycle stage to another.

One of the reasons that LCA has been successful at YVW is that it aligns with the professional background of the employees. One manager summed this up as:

> *We're a bunch of engineers and accountants and economists with this scientific training and all that sort of hoo-hah. We were looking for something that would give us something a bit more quantifiable in terms of an analysis. Upon adopting life cycle assessment we're now able to quantify the amount of greenhouse gases emitted, the amount of nutrients discharged, and the other impacts that our servicing options might have. We can then identify the best option on an environmental basis.*

In simplified change-management terms, its use meant that the wins became visible, motivating employee engagement with the change program.

For example, an energy map was developed to demonstrate the variation in energy intensity across different water and sewer supply zones. The highest energy intensive zone uses about ten times more energy to deliver services than the least intensive zone. This information has helped inform alternative options including providing customers with low flow shower heads and alternative sewer systems.

For example, at Gembrook a pressure sewer system was designed and installed. Although a pressure sewer system is more complex than a gravity-based design, the environmental and community benefits are achieved through less energy-intensive pumping stations and by laying the pipes at a shallower depth than traditional gravity sewers. The Gembrook project provides a tangible example of how cultural change combined with the appropriate sustainability tools can create significantly improved business and sustainability outcomes. Its success has contributed to LCA now being the standard approach that is used to assess environmental impact and to inform key decisions.

In much the same way that the cultural survey tools such as the Organisational Culture Inventory provided data on the culture and supported the establishment of a more open, collaborative and innovative culture, LCA has broadened staff understanding of environmental impact. By quantifying these impacts more effectively and systematically than traditional environmental management tools, it has played an important role in changing the *technical* assumptions that inform significant decisions.

For example, when faced with implementing a sewerage system in suburbs such as Gembrook, the earlier YVW approach, informed by historical technical assumptions that bigger is better and a centralised system is easier and cheaper to maintain, would have been to install sewer pipes which all feed into a centralised system. In the old YVW culture, where new ideas were considered risky and actively discouraged, technical assumptions such as these would be less likely to be scrutinised. Indeed, behaviour that challenges assumptions and the status quo would typically be actively discouraged.

However, with YVW's more constructive culture in which *innovation* was actively encouraged and *environment* was a central part of the organisation's strategic intent, LCA provided a means to quantifiably measure the environmental, social and financial impacts of technical options over their life cycle.

# Collaborative learning through stakeholder engagement

Stakeholder engagement is now widely recognised as a mechanism by which organisations can learn and adapt to sustainability challenges. The culture change at YVW is associated with adoption of this approach as a means of bringing about change. Consistent with the historically risk averse and monopoly culture that had developed in the water sector, there had been significant trepidation when it came to relating with external stakeholders including customers, developers, government and organisations (such as catchment management authorities) that made up the water value chain. There has been a realisation that important sustainability outcomes have been achieved by effectively encouraging innovation and this has led to a significant shift in how YVW collaborates and engages with external stakeholders. YVW personnel can also see the potential for innovative solutions with stakeholders if the same principles are used.

For example, one manager describes an early situation where YVW were approached by a developer that was keen to incorporate recycled water into a development. This situation reflects the beginning of a major shift in YVW's approach to working with stakeholders.

> *Typically our relationship with developers had been quite confrontational. But actually one of the key things that we did with a particular developer was to establish a partnership arrangement where there is open discussion and we get insights on what's driving the thinking of each party in the relationship. That was the first time that we worked together and we've been working that way ever since. Oh I guess that's a key way that our culture has changed. We now actually work with partners for a collective gain, and it also helps us collaborate on and work in areas outside of our expertise, for example harvesting stormwater for drinking water delivers business benefits to us and environmental benefits to local waterways.*

YVW has also incorporated stakeholder views into a sustainability assessment framework that builds on the technical aspects highlighted by LCA by allowing environmental, financial and social costs and benefits to be evaluated in an integrated way. As one senior manager describes it:

> *The sustainability assessment framework allows us to assess different options on a triple bottom line basis, so as well as the environmental impacts and the NPV (net present value) calculation we look at the impact on culture and on the customer as well.*

Another manager describes the benefits:

> *I guess that's the exciting part of what we've done in the last few years is bring together the major players in the urban water cycle, and get highest level agreement that there's ... a prize for us all, and then get a philosophical agreement to work together and then stepping it down into working groups to actually find a different solution.*

# Linking culture, strategy and mechanisms for learning and change

YVW is widely recognised as a sustainability leader in the water industry. Its success is the outcome of a decade-long and continuing focus on organisational learning and change.

There are many dimensions to YVW's approach to sustainability and organisational change. This case study has described YVW's approach and highlighted the important interconnection between three critical aspects of the change process:

1. **Organisational culture**. Through measurement, consultation and the effective implementation of a range of new management practices, the culture is more open, there is a high level of collaboration within and outside the organisation, and innovation is actively encouraged. This is essential for sustainability since it requires new attitudes, thinking and applications in order to achieve improved social and environmental outcomes. In particular, the downplaying of hierarchy and status combined with appreciation rather than punishment of the open expression of ideas and feelings has led to an upward flow of innovative ideas within the organisation.

2. **Integration of *environment* as a strategic issue**. Since environment is communicated as an issue of strategic importance, staff are actively encouraged to integrate environmental considerations into the work that they do. Essentially, discretionary projects that deliver on all strategic elements will have a greater chance of success than those that do not – and to do this requires environmental deliverables. The incorporation of environmental goals into key performance indicators has reinforced the view that YVW is serious about its intent in this area.

3. **Sustainability tools and mechanisms**. TNS, LCA and the incorporation of internal and external stakeholder perspectives into key decisions has improved the way in which sustainability impacts are measured and understood. This has supported a more systematic and informed approach to integrating environmental considerations into decision making.

## Organisational learning

To deliver on key environmental strategies, YVW needed, among many other things, to encourage its staff and key stakeholders to become passionate about the environment. The organisation needed an experiential process where people could see the link between personal and organisational decision-making, behavioural change and YVW's impact on the environment.

The key components of the education program were:

- a program of speakers delivering inspirational environmental messages that encouraged innovation and engagement. These speakers included international and local experts, including YVW staff

- an Environmental Sustainability Training Program

- an intranet site with supporting information and tools

- 'The Passion Index', a survey designed to help staff assess their own personal impact on the environment

- a World Environment Day Expo of activities, which invited staff to travel to work in an environmentally friendly way, create a huge environmental map of Melbourne, measure their ecological footprint, view educational videos and other stimulus material, and receive an indigenous plant for their efforts.

The program design was developed and refined with input from:

- the YVW executive team

- the 'Sustainability Circle' (an internal environmental leadership group which met monthly)

- a team of environmental 'champions' from across the business

- the Environmental Sustainability Advisory Committee (ESAC), which is made up of leading environmental experts from around Australia that meets on a regular basis and is encouraged to propose new and innovative ideas.

The combined influence of the range of measures implemented at YVW can be seen in the changing attitudes and assumptions that inform decision making on both small and large-projects. Some of the more significant shifts are summarised in Table 13.1.

**Table 13.1** Shifting assumptions at YVW towards improved decision-making for sustainability

| Old assumptions | New understanding |
|---|---|
| Environment and business are a trade-off | Environment integrates with strategy |
| Big centralised infrastructure systems are best | Decentralised systems can be more cost-effective and reduce environmental impact |
| Major environmental impact is water | Energy AND water are the major contributors to environmental impact |
| Customers only want cheap water | Customers will pay to reduce their carbon footprint |
| Stakeholder collaboration is 'risky' | Broadening stakeholder engagement can create mutual benefit and reduce environmental impact |

This case analysis of the learning and change for sustainability that occurred at YVW leads to some suggestions for information gathering around each of these three critical aspects of the change process, which other organisations can employ in preparing for such an approach (Figure 13.2).

**Figure 13.2** Preparing the ground for change towards sustainability

## Culture:

- What are the characteristics of the current culture?

- What aspects of culture support and hinder progress towards sustainability?

- Is there a cultural change program currently in place?

- Could sustainability initiatives be more closely connected with ongoing efforts designed to modify the existing organisational culture?

## Integration:

- What are the organisation's current strategic issues and challenges?

- What priority is currently placed on sustainability?

- Are there further opportunities to reframe sustainability issues in terms of business value or as a means of addressing some of the existing strategic challenges faced within the organisation?

## Sustainability tools and mechanisms:

- What tools and organisational processes are most suited to the culture and sustainability challenges of the organisation?

- What has or hasn't worked previously? Why?

- Are there opportunities to trial new approaches on current or future projects?

# Conclusion

For managers and teams charged with the challenge of implementing sustainability initiatives in other organisations, the level of resourcing, leadership and support that have been demonstrated in YVW's case may not be available. However, this case study demonstrates the importance of careful consideration by managers in creating an open and supportive culture of innovation, of the necessity of integrating sustainability into strategic planning and ongoing operations, of finding and implementing appropriate sustainability tools as well as focusing on both employee and stakeholder engagement. The case also illustrates the necessity of serious long-term organisational commitment to working towards sustainability rather than the fleeting adoption of short-term fixes or fads.

# Appendix 13.1: Yarra Valley Water awards

## 2010–2011

- Runner up in the International Water Association's Sustainability Specialist Group Prize for Research Excellence. This prize recognises excellence in scientific research relating to sustainable urban water management and attracts entries from leading organisations around the world.

- Winner of the Culture Transformation Sustainability Achievement Award at the Australian Conference on Culture and Leadership. The award recognises the efforts made to effect cultural change and maintain a strong constructive culture that supports innovation and the organisation's vision for the future.

## 2009–2010

- High Commendation in the Management and Initiatives category of National Stormwater Excellence Awards for the Kalkallo Harvesting and Reuse Project.

- Stormwater Industry Association of Victoria Excellence Award for the planned Kalkallo Stormwater Harvesting and Reuse Project. The award was given in the Master-Planning and Design category and recognised the work done in integrating stormwater design principles into the provision of YVW water, sewer and recycled water services.

- Finalist in two categories of the Banksia Environmental Awards; the Water Award and the Large Business Sustainability Award. The Banksia Environmental Awards are regarded as the most prestigious environmental awards in Australia.

## 2008–2009

- Winner of the Premier's Sustainability Award for Large Business for a research project to compare alternative service options such as recycled water, rainwater tanks and greywater recycling, with traditional centralised servicing in new developments.

- Sustainability Specialist Group Prize for Research Excellence (runner up) by the International Water Association. The prize recognises excellence in scientific research relating to sustainable urban water management, and attracts entries from leading organisations all over the world.

## 2007–2008

- Finalist in the Greenhouse Challenge Plus 2007 awards for Mitcham Green Office Strategy, which delivered substantial reductions in the use of energy, water and waste generated.

## 2006–2007

- Winner of International Water Association's Best Customer Account Award in the Marketing and Communications category.

- Finalist in the United Nations Association of Australia World Environment Day Award in the Triple Bottom Line category for the Gembrook Sewerage Servicing Strategy Case Study.

## 2005–2006

- Received the Prime Minister's Award for Excellence in Community Business Partnerships (Large Business Award Victoria) for YVW partnership with Kildonan Child and Family Services.

- In September 2005, Yarra Valley Water won the MIS (Managing Information Strategies) Innovation Award for Best Governance Initiative, which recognises excellence in the use of information technology. The award was given for YVW work in the upgrading of our IT systems.

## 2004–2005

- Winners of State and National Service Excellence awards for the Government sector from the Customer Service Institute of Australia,

- Won the 2004 award for 'Best Change Management' from Australian Human Resources Magazine.

- Australasian Risk Management Awards, securing two out of the five award categories, namely:

  - The Award for Best Implementation of Risk Strategy

  - The Award for Best IT Security Strategy.

- Our Risk Manager, Frank Portelli was also a finalist in the Risk Manager of the Year category.

## 2003–2004

- YVW the Customer Service Institute of Australia (CSIA) Victorian and National Awards for Australian Service Excellence in the Government category.

- Finalist in Premier's Awards for Business Sustainability for our savewater!™ Efficiency Service.

- In October 2003, YVW received the Large Business Encouragement Award for Victoria at the Prime Minister's Awards for Excellence in Community Business Partnerships.

- Won Victorian Call Centre of the Year and Victorian Teleprofessional of the Year. Yarra Valley Water was the winner of the Geospatial Information Technology Association (GITA) Award for Excellence.

- Won awards for IT processes including the Government Technology Productivity Award and Intergraph 100% Club Award.

## 2002–2003

- Recognised by the **Committee for Melbourne** in 2002 with one of their leading awards in sustainability.

- YVW involvement with the **Young Achievement Australia** (YAA) program resulted in winning the scheme's ultimate award, **Company of the Year**, in November 2002.

# References

Jones, Q, Dunphy, D *et al*. 2006, 'In Great Company - Unlocking the Secrets of Cultural Transformation', *Human Synergistics*, Sydney, Australia.

Kelly, T 2006, 'Practical application of the Triple Bottom Line in a water utility', International Water Association, Beijing.

Pamminger, F & Crawford, J 2006, 'Every journey starts with a single step - Yarra Valley Water's Journey Towards Environmental Sustainability', 12th ANZSYS conference - Sustaining our Social and Natural Capital, Katoomba NSW Australia.

Pamminger, F & Kenway, S 2008, 'Urban metabolism - improving the sustainability of urban water systems', *Journal of the Australian Water Association*. 25(1): 28-29.

Pamminger, F & Narangala, R 2009, 'LCA's evolution and integration into sustainable business decisions: a water company's perspective', Sixth Australian Conference on Life Cycle Assessment, Melbourne, Victoria, Australia.

Robèrt, KH, Schmidt-Bleek, B *et al*. 2002, 'Strategic sustainable development -- selection, design and synergies of applied tools', *Journal of Cleaner Production*. 10(3): 197-214.

# INDEX

## A

adaptive management, 17, 19, 22
advertising, 116, 117
alliance performance, 96
assessment processes, environmental, 153
audits, 36, 53, 102, 104

## B

best practice, 38, 59, 94, 121, 124, 130, 135
bottom-up initiatives, 59
business goals, 76
business model, 2, 6, 7, 38, 80, 90, 108, 110, 111
business strategy, 2, 32, 81, 83, 94, 128
business systems, 121

## C

capital drain, 4
capital investment, 85, 90, 124
carbon inventory, 102
champion effectiveness, 21
champions of innovation, 11, 21, 23
change agents, 10, 11, 14, 20, 21, 23
*Child Labour Deterrence Act*, 70
climate change, xxvii, 9, 93, 99, 101, 139, 142, 152
code of conduct, 49, 52, 64
collective bargaining, 66, 72
community engagement, 3, 5, 6, 58
community projects, 4, 53, 80
competitive advantage, 5, 44, 45, 61, 89, 122
components, remanufacturing, 34
continuous improvement, 94, 129
corporate responsibility, 94
corporate sustainability, 20, 40, 60, 97, 138, 139, 140, 141, 142, 143, 144, 147

cross-boundary teams, 21
cultural change program, 149
culture change, 39, 155

## D

data analysis, 3, 14, 77
development assessment process, 17, 18, 19, 21, 23
development initiatives, 7, 21
development planning cycle, 133
diversity, 35, 95, 142
Dow Jones Sustainability Index, 138, 139
drainage, 9, 12

## E

ecological sustainability, 39, 60, 62, 139
economy of scale, 118
embedding sustainability, 83
enabling contextual factors, 10
enabling factors, 17
enabling leadership, 15, 19, 21, 22, 23
environmental costs, 28
environmental education program, 152
environmental responsibilities, 49, 152
ethical trading, 64
executive champions, 11, 13, 18, 22

## F

feedback, 34, 46, 52, 93, 109, 128, 129, 130, 132, 134
feedback, 360 degree, 132
financial reward systems, 34
financial structures, 122
forced labour, 64

## G

GHG emissions, 87, 100, 103, 104, 105
GHG measurement, 105
global financial crisis, 7, 77, 88, 89, 122, 126, 136